Bioluminescent Marine Plankton

Authored by

Ramasamy Santhanam

Fisheries College and Research Institute
Tamil Nadu Veterinary and Animal Sciences University
Thoothukudi
India

Bioluminescent Marine Plankton

Author: Ramasamy Santhanam

ISBN (Online): 978-981-5050-20-2

ISBN (Print): 978-981-5050-21-9

ISBN (Paperback): 978-981-5050-22-6

need for a court order if at any point you breach any terms of this License Agreement. In no event will any delay or failure by Bentham Science Publishers in enforcing your compliance with this License Agreement constitute a waiver of any of its rights.

3. You acknowledge that you have read this License Agreement, and agree to be bound by its terms and conditions. To the extent that any other terms and conditions presented on any website of Bentham Science Publishers conflict with, or are inconsistent with, the terms and conditions set out in this License Agreement, you acknowledge that the terms and conditions set out in this License Agreement shall prevail.

Bentham Science Publishers Pte. Ltd.
80 Robinson Road #02-00
Singapore 068898
Singapore
Email: subscriptions@benthamscience.net

BENTHAM SCIENCE

CONTENTS

PREFACE

Bioluminescence, the "cold living light" or the "cold fire of the sea," is extremely common in all oceans at all depths. However, this phenomenon is nearly absent in freshwater, with the exception of a freshwater limpet. More than 75% of deep-sea creatures have been reported to produce their own light. The luminescent marine plankton such as dinoflagellate, radiolarians, jellyfish, comb jellies, annelids, copepods, ostracods, mysids, amphipods, euphausiids, and tunicates form an important component in the marine food chain. Research on luminescent marine plankton is gaining momentum now-a-days owing to its importance in human health. The glowing Green Fluorescent Protein (GFP) extracted from the North Pacific jellyfish, *Aqueorea victoria* (for which the Japanese biologist, Osamu Shimonmura won the Noble Prize in Chemistry in 2008) has helped shed light on key processes such as the spread of cancer, the development of brain cells, the growth of bacteria, damage to cells by Alzheimer's disease, and the development of insulin-producing cells in the pancreas. Furthermore, GFP has been "a guiding star for biochemists, biologists, medical scientists and other researchers" besides serving as an indispensable tool in cellular research and medicine. Recent research findings have also shown that the natural products of the bioluminescent marine plankton could be of great use in therapeutical and biotechnological applications. Further application of "bioluminescence imaging" has grown tremendously in the past decade, and it has significantly contributed to the core conceptual advances in biomedical research. This technology has provided valuable means for monitoring different biological processes for immunology, oncology, virology, and neuroscience. Bioluminescence imaging has also been successfully used to monitor infections caused by various microorganisms, particularly bacteria.

Though a few books are presently available on bioluminescence, a comprehensive volume dealing with the "Bioluminescent Marine Plankton " has not so far been published. The first of its kind, this publication would answer this long-felt need. It deals with the chemistry of bioluminescence, types of bioluminescent displays, distribution of bioluminescence among marine plankton, ecological functions and uses/applications of planktonic bioluminescence; and the biology and ecology of about 200 species luminescent marine plankton of the different seas. It is hoped that the present publication, when brought out, would be of great use as a standard text-cum-reference for teachers, students, and researchers of various disciplines such as Marine Biology, Fisheries Science, and Environmental Sciences; and as a valuable reference for libraries of colleges and universities.

CONSENT FOR PUBLICATION

Not applicable.

CONFLICT OF INTEREST

The author declares no conflict of interest, financial or otherwise.

ACKNOWLEDGEMENTS

I am highly indebted to Dr. K.Venkataramanujam, former Dean of Fisheries College and Research Institute, and the Tamil Nadu Veterinary and Animal Sciences University, in

Thoothukudi, India, for his valued comments and suggestions on the manuscript. I sincerely thank all my international friends who were kind enough to share their images for the present publication. The services viz. photography and secretarial assistance rendered by Mrs.Albin Panimalar Ramesh are also gratefully acknowledged.

Ramasamy Santhanam
Fisheries College and Research Institute
Tamil Nadu Veterinary and Animal Sciences University
Thoothukudi
India

CHAPTER 1

Introduction

Abstract: This chapter deals with the different types of luminescence: chemiluminescence, photoluminescence, fluorescence, phosphorescence, and bioluminescence; types of bioluminescence *viz.* extra cellular luminescence, intrinsic luminescence and extrinsic luminescence (bacterial luminescence); distribution of bioluminescence in different groups of marine plankton such as dinoflagellates, crustaceans, cnidarians, ctenophores and tunicates; biological functions of bioluminescence; interactions among luminescent marine zooplankton and fish; and commercial and therapeutic applications of bioluminescence.

Keywords: Applications of bioluminescence, Bioluminescent marine life, Cold living light, Extrinsic luminescence, Intrinsic luminescence.

The Greeks and Romans were the first to mention bioluminescent organisms, and Aristotle (384 – 332 BC) discovered self-luminosity in 180 marine species [1]. Bioluminescence, the cold living light, is a visible light produced by organisms. It is one form of chemiluminescence generated by a chemical reaction. Unlike fluorescence and phosphorescence, bioluminescence reactions do not need the initial absorption of sunlight by a molecule to emit light. A wide range of colours characterize bioluminescence. For example, it is blue in jellyfish, dinoflagellates, and ostracods; blue-green in bacteria and the limpet-like snail, *Latia*; and green-red in fireflies and railroad worms [2]. Some species of fungi are known to emit light continuously, and its glow is called foxfire. Most organisms, however, flash for periods of less than a second to about 10 seconds and these flashes occur in specific spots, such as the dots on a molluscan squid [3]. Bioluminescence has been reported to be very common in the water column of the ocean, less common on coral reefs and other places near the shore, rare on land, and nearly non-existent in freshwater (with the exception of a limpet-like freshwater snail, *Latia neritoides,* and a land snail, *Quantula striata*). The bioluminescence of the sea otherwise known as "cold fire of the sea" is present at all depths. It is estimated that 76% of the animals of the seas and oceans are bioluminescent. About 800 genera representing 13 phyla of marine animals are believed to be bioluminescent [4]. It is worth- mentioning here that for most of the well-known ocean glows, the dinoflagellate blooms are largely responsible.

LUMINESCENCE AND ITS TYPES

Luminescence is the process of giving off light and is defined as any emission of light from a substance that does not arise from heating. There are many types of luminescence, *viz.* "Chemiluminescence" where a light emission is initiated by a chemical reaction; "photoluminescence " is the emission of light from a material following the absorption of light; "fluorescence" is prompt photoluminescence that occurs very shortly after photoexcitation of a substance; "phosphorescence" is long-lived photoluminescence that continues long after the photoexcitation has ceased; and "bioluminescence" is the production and emission of light by a living organism and it is otherwise known as "cold living light."

COLOURS OF BIOLUMINESCENT LIGHT

While most land organisms, including fireflies, glow in the yellow spectrum, almost all marine bioluminescence is blue due to two related reasons. Firstly, the blue-green light (wavelength around 470 nm) transmits furthest in water. Underwater photos are usually blue because water absorbs red light quickly as one descends. Secondly, most marine organisms are sensitive only to blue light, and they lack visual pigments that can absorb longer (yellow, red) or shorter (indigo, ultraviolet) wavelengths [5].

LIGHT EMISSION IN MARINE ANIMALS

While the bioluminescence of marine animals is invariably blue, the colour of the light can range from nearly violet to green-yellow (and very occasionally red), emitted in three different ways. In some species, the light is actually vomited from the animals. In other animals, the light is emitted by specialized cells called photocytes,sometimes grouped into lensed structures called photophores. On the other hand, some animals may have colonies of bioluminescent bacteria which glow continuously [6].

Types of Bioluminescence

Based on the source of illumination in marine animals, the luminescence may be classified as follows:

i. **Extracellular Luminescence**: In this type of luminescence, the light is generated by luminous secretion from the glandular tissues of animals. Extracellular luminescent organs are found in a very limited species of fish. For example, certain fish like rat tails emit light by secreting extracellular slime. These fish possess special glands near their anus, which secrete luminous slime sufficiently.

ii. **Intrinsic Luminescence (Intracellular Luminescence)**: In this type, light is produced intracellularly and the light is emitted by special cells called photocytes which form light producing photophores developed from the epidermis. This type of luminescence is mainly seen in teleost fish of the families, such as Sternoptychidae, Myctophidae, Halosauridae, Stomiatidae, Brotulidae, Lophiidae, and Zoarcidae.

iii. **Extrinsic Luminescence (Bacterial Luminescence)**: In this type, symbiotic bacteria present in the photophores or luminous cells (photocytes) discharge light. The bacterial genera, Photobacterium and Achromobacterium, have been reported to contribute much to this type of luminescence. These bacterial species commonly found in dead fish or spoiling meat, have been isolated and grown in cultures.

Types of Bioluminescent Displays

Based on the appearance, the bioluminescent displays may be classified into three types *viz.* sheet type, spark type, and gloving-ball or globe type.

i. **Sheet-type Display**: This is the most common type in coastal waters and is caused by dinoflagellates or bacteria. It is also known as spilled or "milky" bioluminescence. During the formation of this type, the seawater is cloudy and may appear dully luminescent. The colour of the water is usually green or blue, and in many displays, it may also appear white when the organisms are present in great numbers *i.e.*, during the bloom formation. Apart from the microscopic organisms, dense and extensive concentrations of large tunicate organisms, such as luminescent Pyrosoma (giving a flashing appearance during lower concentrations) or luminescent euphausiid species Nyctiphanes norvegica may also yield sheet-type display. The display of this euphausiid species was associated with large spots and long bands of milk-white water [7].

ii. **Spark-type Display:** This type of display is largely due to the appearance of large numbers of luminous euphausiids or copepods. This display occurs most often in colder waters and only when the waters are disturbed. The luminescence colour during this display is brilliant blue or white [7].

iii. **Glowing-ball- or Globe-type Display**: This display is normally observed in the warmer waters of the world. During this condition, the ocean may appear as full of balls or discs of light which may be flashing brightly when they are disturbed or dimming after the stoppage of the initial stimulus. Depending on the size of the luminescent organisms, the flashes of light may range in size from a few centimeters to a few meters in diameter. The colour of the light during this display is normally blue or green, and rarely it may be white, yellow, orange, or red. The light so emitted may rarely be continuous (Staples,1966). Combinations of either two or all three types of displays have also been reported. Apart from these displays, exotic light formations like "phosphorescent wheels," undulating waves of light, and bubbles of light have also been reported due to large concentrations of luminescent organisms [8].

iv. **Other Bioluminescence:** Marine organisms normally luminesce when they are disturbed. However, changes in the marine environment, such as a drop in salinity, may force bioluminescent algae to glow, for instance. These living lanterns are often seen as spots of pink or green in the dark ocean. "Milky seas" are yet another example of bioluminescence. Unlike the bioluminescent algae, which emit light when their environment is disturbed, these milky seas yield continuous glow which may be bright and large enough to be seen even from satellites in orbit above the earth. "Milky seas" are believed to be produced by the millions of bioluminescent bacteria present on the surface of the ocean. Satellite imagery of milky seas has been captured in tropical waters of the Indian Ocean [9].

BIOLUMINESCENT EMISSION SPECTRA OF MARINE ANIMALS

Light production associated with bioluminescence in marine animals has a significant range of emission patterns, *viz.* continuous glow (single flashes of light) common in the phytoplanktonic dinoflagellates, or repetitive pulse patterns that are often species specific. Visible light (*i.e.* VIBGYOR- V to R) has wavelengths in the range of 400–700 nm *i.e.*, between the infrared (with longer wavelengths) and the ultraviolet (with shorter wavelengths). Most non-marine organisms are generally at longer wavelengths (480-620 nm). For example, millipedes emit indigo (and coleopteran beetles emit reddish-orange light. On the other hand, in marine organisms, the emission maxima are clustered in the range of 450-500 nm, though maxima from 395-545 nm have been recorded. Marine species found in the pelagic environment are mostly blue-emitting, with a relative increase in green-emitting species in the benthic environment [10]. Although, it is reported that the majority of luminescent marine organisms emit blue light (410–550 nm), a change from violet and blue (420–500 nm) in the deep sea to blue-green (460–520 nm) in shallow waters is common [11]. Further, the colour of the bioluminescent light is largely dependent on factors such as the luciferins and luciferases, which are involved in the bioluminescent reaction.

DISTRIBUTION OF BIOLUMINESCENCE IN MARINE ORGANISMS

Bioluminescence exhibits a diversity of organisms from bacteria to fish, and it has been shown that 76% of the marine animals are bioluminescent. The percentage of bioluminescent marine animals is remarkably uniform over depth. Moreover, the proportion of bioluminescent and non-bioluminescent animals within taxonomic groups changes with depth, especially for taxonomic groups such as

Ctenophora, Scyphozoa, Chaetognatha, and Crustacea [12]. The luminescent marine invertebrates include protozoans (*viz.* dinoflagellates and radiolarians), sponges, jellyfish, comb jellies, sea pens, worms, copepods, ostracods, mysids, amphipods, euphausiids, shrimps, squids, and echinoderms; and among the luminescent vertebrates, the tunicates and fish possess several species. It is worth mentioning here that more than 97% of Cnidarians are bioluminescent; among 20,000 known species of fish, about 1500 (8%) species are luminescent. While the majority of deep-sea bony fish (about 70%) are luminous, only a small fraction of deep-sea elasmobranchs (about 6%) are endowed with bioluminescence.

DISTRIBUTION OF BIOLUMINESCENCE IN MARINE PLANKTON

A total of 556 species of marine zooplankton have been reported to display bioluminescence. Among them, the crustaceans dominated with 283 species followed by cnidarians (92), protozoans (90), ctenophores (45), tunicates (22), annelids (15), echinoderms (4), chaetognaths (3), and gastropod molluscs (2) [13].

Dinoflagellates: Marine planktonic organisms which emit blue and green light are mainly responsible for bioluminescence in the sea. Among these organisms, the dinoflagellates of the genus *Noctiluca* assume greater significance, and they are often responsible for strong displays of light owing to their prodigious numbers. They colour the seawater pink or red by day. *Noctiluca* is particularly abundant in coastal waters, and at night it gives a rather brilliant greenish glow to the water when agitated. Other dinoflagellates of equal importance include the species of *Phaeocystis, Ceratium, Peridinium, and Gonyaulax* [14]. The genus *Gonyaulax* also is one of the prime causes of "red tide."

Crustaceans: Among the luminescent crustaceans, ostracods, copepods, and euphausiids are important and most of their displays are seen in colder waters and rarely in tropical waters. The light emitted by these animals appears to twinkle at a distance because of each individual's abrupt flashing, usually in blue or green [7].

Cnidarians and Ctenophores: Among the cnidarians (=coelenterates), the luminescent jellyfish (scyphozoan medusae) have been reported to cause many bright displays which may cover a large area. One of the most spectacular forms of the Cnidaria is the large luminescent medusa *Pelagia noctiluca*. When it is touched lightly, the whole surface of the organism starts to luminesce, first at the point of contact and then spreading out to its umbrella and tentacles. It is worth mentioning here that all the ctenophores (comb jellies) are luminescent, giving off a greenish glow [7].

Tunicates: Among the transparent tunicates, the colonial species of *Pyrosoma* are largely responsible for some of the prominent displays, especially in the warmer waters in the seas and oceans. The slightest touch at one end of the colony can cause blue light in these species. Various colours of *Pyrosoma* luminescence, such as red, orange, yellow, and white light, have been reported. However, the light normally given off in the sea by these organisms is bluish-green or green. In colder waters, the luminescent salps are often in great abundance, and they may be present as individuals or in great chain-like aggregations. The species of Salpa emit blue or green light [7].

Others: Among the other important marine bioluminescent groups of organisms are the luminescent annelid worms, *Odontosyllis*. Similarly, the deep-sea squid *Watasenia scintillans* (Japanese firefly squid), has been reported to congregate in large numbers on the surface, especially during the spring, giving displays. The microscopic radiolarians may rarely impart a weak luminescence in the ocean. The bioluminescent marine animal groups, theirnumber of species, and percentage contribution are given in Table **1**.

BIOLOGICAL FUNCTIONS OF BIOLUMINESCENCE

Bioluminescence of marine animals assumes several important functions including predation, defense against predators, and reproduction. In other words, this adaptation can help animal survival in at least three critical ways *viz.* it can help in locating food, either by means of built-in headlights or by the use of its glowing lures; it can function as a defense against predators; and it can be used to attract its mate by means of species-specific spatial patterns of light emission.

INTERACTIONS AMONG LUMINESCENT MARINE ZOOPLANKTON AND FISH

Fish–luminescent marine plankton interactions such as predation by fishes on luminescent plankton, competition between luminescent zooplankton and fish are complex. However, these interactions have not so far been convincingly demonstrated. The bright coloration of animals is believed to be toxicity or unpalatability and this phenomenon is also quite valid for many gelatinous planktonic marine organisms such as cnidarian jellyfish and ctenophores; and tunicate pyrosomes. The coloration of such bioluminescent zooplanktonic organisms has been reported to Function this way. As these jellyfish are fragile and potentially deadly, they make use of their bright coloration as an adaptation to avoid physical encounters with their predators such as fish.

It is reported that the luminescent Jellyfish species *Pelagia noctiluca* which forms very large population outbreaks with millions of individuals that hinders the physiology of fish including their growth and reproduction. Further, the ctenophore *Mnemiopsis leidyi* was found to be responsible for the collapse of the anchovy fishery in the Black Sea. Furthermore, potentially harmful species such as the scyphomedusae *Pelagia noctiluca* and the siphonophore *Muggiaea atlantica* have affected the fish farming practices in the North Atlantic region [15].

In coral and seagrass beds, the luminescent jellyfish have also been reported to provide shelter for the juvenile carangid fishes beneath their sub-umbrella. It is suggested to be an act of commensalism where the tentacles of the jellyfish served as a protective shield for the juvenile fishes against their predators [16].

APPLICATIONS OF BIOLUMINESCENCE

Research on the bioluminescence of marine invertebrates and fishes is gaining momentum owing to its importance in human health.

Green Fluorescent Protein (GFP)

The glowing Green Fluorescent Protein (GFP) extracted from the North Pacific jellyfish, *Aqueorea victoria* (for which the Japanese biologist Osamu Shimonmura won the Noble Prize in Chemistry in 2008) has helped shed light on key processes such as the spread of cancer, the development of brain cells, the growth of bacteria, damage to cells by Alzheimer's disease, and the development of insulin-producing cells in the pancreas. Further, GFP has been "a guiding star for biochemists, biologists, medical scientists and other researchers" besides serving as an indispensable tool in cellular research and medicine. An American Biotechnology company has now generated a fluorescent mouse to observe cancer's growth. Similarly, a research team in England has produced fluorescent-green testicles on male mosquitoes, for their separation from females to reduce the spread of malaria [17]. The calcium-sensitive photoprotein aequorin and green fluorescent protein (GFP) and their derivatives are presently used for a wide range of applications, including subcellular calcium imaging, cell lineage tracing and gene regulation analysis.

Cypridina System

Among other applications, the Cypridina system has been widely used in bioimaging, in studies of circadian rhythms, and in immunoassays [18].

Bioluminescence Imaging

Recent research findings have also shown that the natural products of the bioluminescent marine invertebrates could be of great use in therapeutical and biotechnological applications. For example, the application of "Bioluminescence imaging" has significantly contributed to the advancement of biomedical research. This technology has helped considerably in the monitoring of different biological processes for immunology, oncology, virology and neuroscience. Biolumines-cence imaging has also been successfully used to monitor infections caused by the different species of pathogenic bacteria [17].

Bioluminescence and People

Certain other uses of bioluminescence of marine animals are in experimental stage. For example bioluminescent trees could help light city streets and highways. If this is materialized, the need for electricity could be considerably reduced. Bioluminescent crops and other plants could be made to emit light when they are in need of water or other nutrients, or when they are ready to be harvested. This would considerably reduce the costs of agriculture. Further the importance of bacterial bioluminescence is well understood in recent days [19]. The luminescence of marine bacteria and planktonic animals possess commercial applications *viz.* source of electricity, detection of toxins, bio-imaging in treatments, navigation aid, and watering plants [20] as shown in Table **2**.

CONCLUSION

Bioluminescence is a fascinating aspect possessed by many of the marine creatures including plankton living in our oceans. While scientists have been aware of this ability and its mechanism for centuries, we are still far from understanding everything about bioluminescence in general and marine plankton in particular. Indeed, researchers have not discovered all the reasons why marine plankton are bioluminescent. Also, the chemical reaction which creates bioluminescence, while understood for some marine animals, remains secret for many planktonic animals, such as some planktonic worms and molluscs. It is therefore important for scientists to keep studying bioluminescence for the benefit of humans.

CHAPTER 2

Chemical Mechanism of Bioluminescence

Abstract: This chapter deals with the basic bioluminescence reactions, emission maxima, and colour of light in pelagic and deep-sea bioluminescent organisms, luciferins of planktonic organisms, types of intrinsic bioluminescence such as coelenterazine-based light production and cypridina luciferin-based light production, and extrinsic bioluminescence.

Keywords: Cypridina Luciferin, Emission Maxima, Extrinsic Bioluminescence, Intrinsic Bioluminescence Coelenterazine, Luciferin.

INTRODUCTION

In the seas and oceans, there is an amazing diversity of organisms that emit light, wherephytoplankton and zooplankton play a significant role in the ecology and food chain of these marine environments. While the bioluminescence chemistryis diverse an enzyme-mediated reaction between molecular oxygen and an organic substrate is vital in light emission. The different components of plankton occupying the surface, are presented in the chemical aspects of a major bioluminescence process.

Bioluminescence is considered a "cold light" (*i.e.* cold living light) which states that only a small percentage of this light contains heat, unlike the light produced by the sun's rays or fire (Tampier, 2017).

Luciferase

(A) Luciferin (substrate) + O_2 ------------------→ Oxyluciferin + CO_2 + hv (light)

(B) Photoprotein + Ca^{2++} ---------------→ Protein- coelenteramide *+ CO_2 = hv (light)

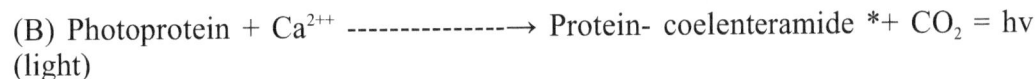

* = Protein-bound oxyluciferin

COLOURS OF BIOLUMINESCENCE

Light at shorter wavelengths, such as blue (400-500 nm) and green (500-600 nm), travels farther down in the ocean (deeper than 100 m). This may be why most marine organisms that emit blue or green lighthide in their surroundings. Further, marine organisms of coastal areas typically produce green light (490-520 nm), whereas the vast majority of pelagic and deep-sea bioluminescent organisms emit blue light with emission maxima (λ_{max}) ranging from 450 to 490 nm [21]. Unlike blue or green light, red light with wavelengths (600-700 nm) are absorbed quickly in the ocean *i.e.*, they travel across the shallower sea but fail to reach the deep-sea zone. Therefore, some deep-sea organisms use red pigmentations on their skin to make them invisible. Further, this red colouration helps deep-sea animals camouflage in the depths where they appear black and disappear into the darkness [22]. The spectral properties in terms of emission maxima of certain marine bioluminescent species are compared with that of terrestrial species in Fig. (**1**).

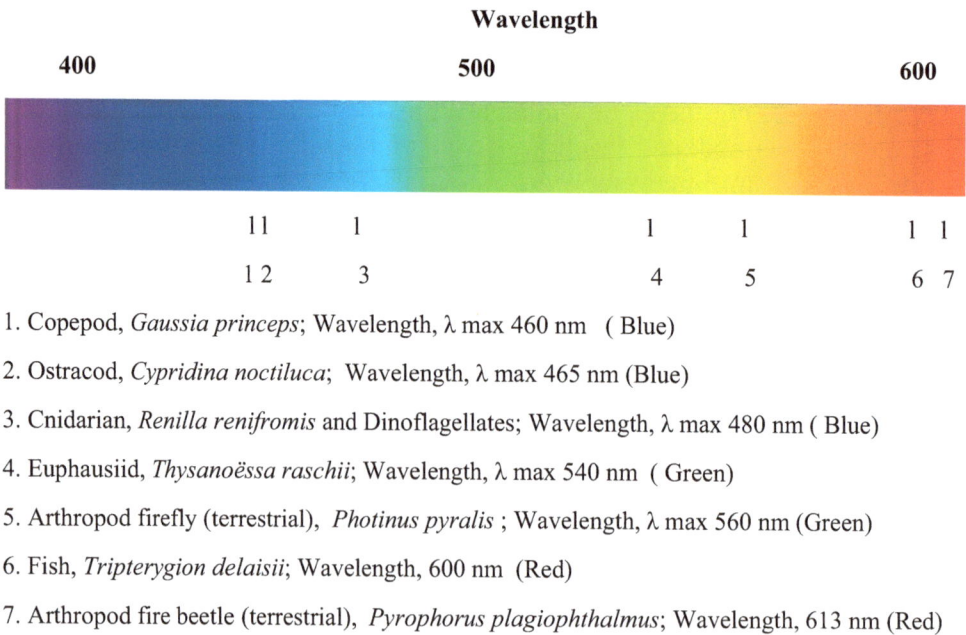

1. Copepod, *Gaussia princeps*; Wavelength, λ max 460 nm   (Blue)

2. Ostracod, *Cypridina noctiluca*; Wavelength, λ max 465 nm (Blue)

3. Cnidarian, *Renilla renifromis* and Dinoflagellates; Wavelength, λ max 480 nm (Blue)

4. Euphausiid, *Thysanoëssa raschii*; Wavelength, λ max 540 nm (Green)

5. Arthropod firefly (terrestrial), *Photinus pyralis* ; Wavelength, λ max 560 nm (Green)

6. Fish, *Tripterygion delaisii*; Wavelength, 600 nm (Red)

7. Arthropod fire beetle (terrestrial), *Pyrophorus plagiophthalmus*; Wavelength, 613 nm (Red)

Fig. (1). Spectral properties of marine and terrestrial animals.

Emission Maxima of Marine Organisms

The values of the emission spectra of protozoans , zooplankton, and fishes measured are given in Table **1**. It is worth mentioning here that with few exceptions, emission maxima of these organisms ranged between 440 and 500 nm, in the blue region of the spectrum Table **1** [23].

Table 1. Emission maxima (λmax) values of marine organisms.

Phyla / Class	λmax (nm) range
Protozoa	443- 458
Cnidaria	444- 488
Ctenophora	478- 496
Annelida	565
Crustacea	444- 492
Mollusca	449 - 514
Tunicata	471, 493
Pisces	477 – 689

BIOLUMINESCENCE IN MARINE PLANKTON

Among all bioluminescent marine planktonic organisms, the cnidarians (formerly coelenterates) have been reported to give the brightest luminescence when they were mechanically stimulated. Further, the hydrozoan medusae and siphonophores of these cnidarians emitted at the shortest wavelengths, with emission maxima between 444 and 466nm. The siphonophore *Arnphicaryon* species showed peak emissions at longer wavelengths (maximum at 487nm; and the scyphozoan medusae had peak emissions at wavelengths between 450 and 480 nm. On the other hand, the longest wavelength emissions for the ctenophores, had an emission maxima between 480 and 490 nm. It is also reported that light emission in cnidarians was intracellular, except for the scyphomedusa, *Chrysaora hysosceles*, which produced a luminescent slime. Besides the siphonophores, other short-wavelength emitters are colonial radiolarians which had an emission maxima ranging from 440 to 460 nm. Euphausiids showed narrow bandwidth and the mean observed emission maximum was 461 nm, with shoulders at about 485 and 505 nm. Copepods and ostracods were found to show secreted bioluminescence with emission maxima ranging from 470 to 490 nm. The pelagic polychaete, *Tomopteris nisseni* was found to produce yellow light of intracellular origin within its parapodia and had maximum emission at 565 nm. In some of these organisms both mechanical and electrical stimulations were found to induce luminescence [23]. The groups of planktonic organisms possessing luciferin are shown in Table **2**.

Luciferins

The first luciferin (from the Latin lucifer, "light-bearer"), the vital evidence for our understanding of today's bioluminescence was isolated by Green and McElroy in 1956 [25]. There are five known distinct chemical classes of luciferins to date, namely, aldehydes, benzothiazoles, imidazolopyrazines, tetrapyrroles, and flavins. An imidazolopyrazine derivative, aptly named "coelenterazine," is the luciferin found in coelenterates and many other marine bioluminescence systems [26]. Coelenterazine was found in luminous organisms from six phyla: Sarcomastigophora (Radiolaria), Cnidaria, Ctenophora, Arthropoda, Mollusca. and Chordata (Pisces). Only in the first three of these were Ca^{2+}-activated photoproteins (luciferases) present [27]. The different types of luciferin and their chemical structures are given below.

Table 2. Planktonic organisms and their best-known luciferins [24].

Group Bacteria Luciferin	Dinoflagellate Luciferin	Coelenterazine	Cypridina Luciferin
Bacteria	+	-	-
Radiolarians	-	-	+
Cnidarians	-	-	+
Ctenophores	-	-	+
Ostracods	-	+	+
Copepods	-	-	+
Euphausiids	-	+	-
Decapods	-	-	+
Chaetognaths	-	-	+
Luciferins	-	-	-

Bacterial Luciferin: Bacterial luciferin (Fig. **2**) is a reduced riboflavin phosphate oxidized in association with a long-chain aldehyde, oxygen, and a luciferase [28].

Fig. (2). Bacterial luciferin.

Dinoflagellate and Krill Luciferins: Dinoflagellate luciferin (Fig. **3**) is believed to be derived from chlorophyll. In the genus *Gonyaulax,* when the pH lowers to about 6, this free luciferin reacts, subsequently producing light. A modified form of dinoflagellate luciferin is also seen in the euphausiid shrimp, krill [29]. Both dinoflagellate and krill luciferins are shown in Fig. (**4**).

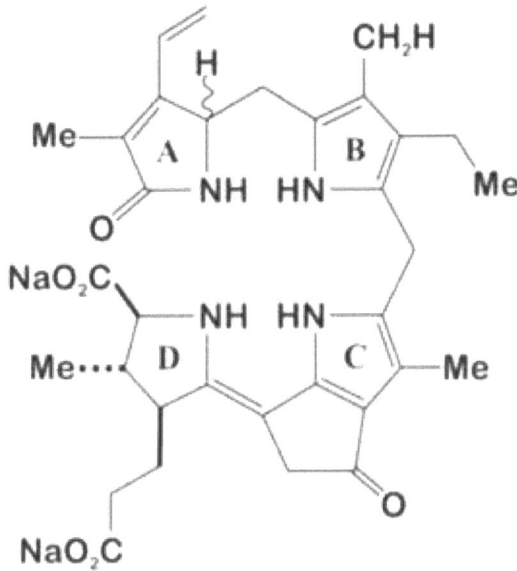

Fig. (3). Dinoflagellate luciferin.

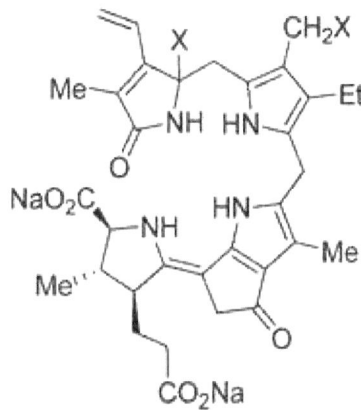

Fig. (4). Dinoflagellate luciferin (X=H); Euphausiid shrimp (krill) luciferin (X=OH).

Vargulin (cypridinluciferin): Vargulin or Cypridina-type luciferin (cypridinidinluciferin) (Fig. **5**) is found in the ostracod ("seed shrimp") *Vargula* and *Cypridina* and is also used by the midshipman fish *Porichthys*. There is a clear dietary link, with fish losing their ability to luminesce until they are fed with luciferin-bearing food. Ostracods have been shown to synthesize this molecule from the amino acids tryptophan, isoleucine, and arginine [30].

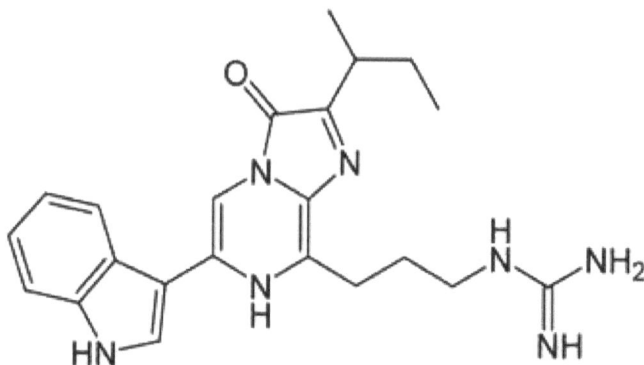

Fig. (5). Vargulin (cypridinluciferin).

Coelenterazine (coelenterate –type luciferin): Among all marine luciferins, coelenterazine (Fig. **6**) is the most popular luciferin. This molecule is the light emitter of the photoprotein "aequorin" and is found in the luminous organisms of six phyla *viz.* Sarcomastigophora (Radiolaria), Cnidaria (= Coelenterata), Ctenophora (= Acnidaria), Arthropoda, Mollusca. Chordata (Pisces) [27]. The natural precursors of coelenterazine have been reported to be the amino acids L-tyrosine and L-phenylalanine [31].

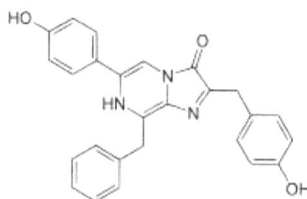

Fig. (6). Coelenterazine.

Luciferases (photoproteins, enzymes): Luciferases consist of 543– 550 amino acid residues and they belong to the adenylate enzyme family [2]. This luciferase is catalyzing the oxidation of the substrate luciferin which is accompanied by the release of energy in the form of light. Although luciferases yield the similar type of light emission in living organisms, their sequences and structures, as well as the mechanisms of the bioluminescent reactions may be completely different in different taxa [32]. Several families of luciferases have been characterized . Aequorin family luciferases are found in cnidarians and ctenophores; Aequorin family photoproteins from hydrozoan species include clytins (or phialidins) from the species of the genus *Clytia* (= *Phialidium*); obelin from *Obelia* spp.; mitrocomin from the species of *Mitrocoma* (=*Halistaura*) and aequorin from the genus Aequorea. The factors associated with the light emission in different groups of luminescent marine animals are given in Table **3**.

Table 3. Bioluminescence in Marine Organisms [2].

Luminous organisms Emission max./nm	Luciferin, Cofactor	Luciferase/kDa
Bacteria (*Photobacterium, Vibrio*) 495–500	$FMNH_2$, RCHO	80
Dinoflagellates (*Lingulodinium, Pyrocystis*) 475–483	Tetrapyrrole, Hþ	130
Cnidarians (*Aequorea, Renilla*) 460–490	Coelenterazine, Ca2þ	25
Molluscs (*Latia*) 536	Enol formate	180
Crustacea (*Vargula; Cypridina*) 465	Imidazopyrazinone	70

Intrinsic Bioluminescence

Bioluminescent systems produce light through the oxygenation of a substrate, called luciferin (lat. lucifer, the light bringer), and an enzyme, luciferase (photoprotein). Bioluminescent reactions vary greatly among organisms but it is typically described as a luciferase catalyzed production of an exciting intermediate from oxygen and luciferin that emits light when returning to its ground state. Many bioluminescence systems may also be triggered by cofactors such as $FMNH_2$, ATP, additional enzymes and intermediate steps for light production. In some bioluminescence systems, special types of luciferases (photoproteins) bind and stabilize the oxygenated luciferin and emit light only in the presence of cations, such as Mg^{2+} or Ca^{2+}. These cations act as a mechanism for the host to precisely control the timing of the light emission. Cofactors in the production of light in respect of *Aequorea* jellyfish and dinoflagellates are calcium ion and proton (Hþ) respectively [2, 11]. It is also worth mentioning here that the variety of colors in bioluminescence is invariably attributed to differences in the structures of luciferase and luciferin [2]. A typical chemiluminescence reaction is given below.

$$L^+ O_2 \xrightarrow[\text{other cofactors}]{\text{Luciferase}} oxy-L^+ \text{lightenergy}$$

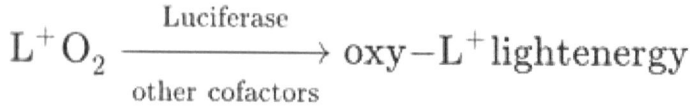

Dinoflagellate Luciferin and Euphausiid (Krill) Luciferin-based Light Production: Both dinoflagellate and krill luciferins are tetrapyrrole-based ones. In dinoflagellates, the production of light occurs in organelles termed scintillations, which contain the luciferin substrate, the luciferase enzyme (LCF) and, in some species, a luciferin binding protein (LBP). The scintillons which are dense vesicles of approximately 0.5–0.9 μm in diameter are abundant in the periphery of the cell especially during the hours of darkness. Flashes of light are primarily produced in these organisms in response to mechanical stimulation due to shear stress which may be [due to contact with grazers or by breaking waves [33].

Coelenterazine-based Light Production: Coelenterazine is a modified tripeptide produced from one phenylalanine and two tyrosine residues. This luciferin assumes greater significance as it serves as a substrate for numerous luciferases of marine organisms. Most of the marine organisms do not synthesize coelenterazine themselves and they obtain it mainly from food. All coelenterazine-dependent bioluminescent systems have been reported to emit blue light, with an emission maxima range of 450–500 nm, and they do not need any cofactors. In some cases, the color of bioluminescence in coelenterazine-based systems is altered by a fluorescent protein that interacts with the luciferase. Other characteristics such as molecular weight, pH-sensitivity, thermostability, and catalysis rates of luciferases differ very much among these coelenterazine-dependent systems [18].

BIOLUMINESCENT REACTION WITH THE LUCIFERIN COELENTERAZINE AND COPEPOD LUCIFERASE

The production of a blue light involving the imidazopyrazinone luciferin, coelenterazine and copepod luciferase (without any additional cofactors) is shown in Fig. (**7**).

Image credit: Svetlana V. Markova, Marina D. Larionova, Eugene S. Vysotski (Reproduced with permission)

Cypridina Luciferin-based Light Production: The luminous ostracods of the family Cypridinidae (commonly called sea fireflies) produce blue light (λmax = 448–463 nm depending on the buffer composition) by an enzyme-catalyzed chemical reaction. This Cypridina luciferin is produced from the amino acids tryptophan, isoleucine, and arginine [18]. In the bioluminescence reaction,

Cypridina luciferin (recently called cypridinid luciferin) is oxidized in the presence of Cypridina luciferase (CLase) (recently called cypridinid luciferase) and molecular oxygen (oxidation step), followed by generation of the oxyluciferin in the excited state (excitation step) and subsequent change to the ground state with light emission (light production step). It is reported that Cypridina luciferin emits spontaneous luminescence in dimethyl sulfoxide (DMSO) without CLase [34]. Further, Cypridina luciferin or its analogs also emitted light more efficiently in diethylene glycol dimethyl ether containing acetate buffer (pH 5.6) or Tris buffer (pH 9.0) containing cetyltrimethylammonium bromide than in DMSO. Furthermore, the mixture of human plasma alpha 1-acid glycoprotein (hAGP) and Cypridina luciferin has also been reported to produce light.

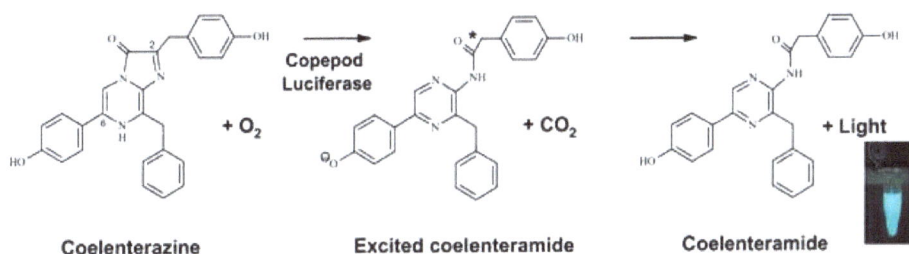

Fig. (7). Bioluminescent reaction with coelenterazine and copepod luciferase. Image credit: Svetlana V. Markova, Marina D. Larionova, Eugene S. Vysotski (Reproduced with permission).

EXTRINSIC BIOLUMINESCENCE

Though all bioluminescent animals possess luciferin, some organisms like the planktonic protozoan dinoflagellates produce their own (intrinsic) and others (squid and fish) absorb bacteria which contain luciferin ((extrinsic) [35]. The luminous bacteria, associated with the marine fauna, are represented by symbiontes which are normally isolated from the light organs, and commensals inhabiting the gastrointestinal tract of host organisms. Five species of such luminous bacteria *viz. Photobacterium phosphoreum, Photobacterium kishitanii, Photobacterium leiognathi, Vibrio harveyi* and *Vibrio fischeri* have been reported. These luminous bacteria are gram-negative bacilli and facultative anaerobes. These halophilous bacteria need only insignificant oxygen concentrations to grow and emit light [36]. The luminescence reaction in bacteria is slightly different from other intrinsic reactions, as it needs luciferase, molecular oxygen, reduced flavin ($FMNH_2$), and a long-chain aldehyde. In this reaction, the luciferase initially forms a complex with $FMNH_2$, which is then oxidized to its peroxide. This complex of luciferase and peroxy $FMNH_2$ reacts with the aldehyde, and it is then broken down to 4a-hydroxy flavin with the production of light and acid. The so formed hydroxy flavin is finally converted to FMN with the loss of H_2O. It is believed that the light yielder in this bacterial luminescence is the luciferase-bound 4a-hydroxy flavin [2]. The biochemical step in this extrinsic (bacterial) luminescence is often linked to the oxidative phosphorylation, in which flavin mononucleotide ($FMNH_2$) reacts with an aldehyde (RCHO) to form a complex (luciferin) which is oxidized to an acid (RCOOH) with the production of light. A typical extrinsic type of bioluminescence reaction is given below

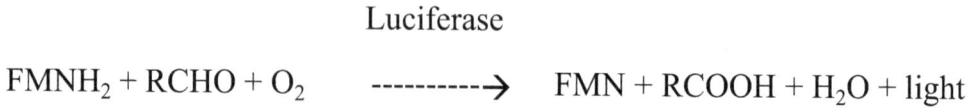

Luciferase

$$FMNH_2 + RCHO + O_2 \quad \text{-----------} \rightarrow \quad FMN + RCOOH + H_2O + light$$

CONCLUSION

Though different types of luciferins *viz.* bacterial, coelenterate, and Cypridina are involved in the chemical reaction of bioluminescence, the Cypridina luciferin-based bioluminescence originated from marine ostracods is regarded as one of the best systems for studying not only bioluminescence mechanisms but also enzyme-substrate interactions and kinetics of enzyme reactions. Many outstanding investigations which have been carried out with other bioluminescent systems would amply tell about the importance of these systems in marine plankton in the near future.

CHAPTER 3

Bioluminescent Marine Dinoflagellates

Abstract: This chapter deals with the identified luminescent species of marine dinoflagellates and their description, emission maxima, total stimulable luminescence (TSL) in observed species of dinoflagellates. and their mechanism of bioluminescence.

Keywords: Dinoflagellates, Emission maxima, HABs, Seafood poisoning, Tetrapyrrole luciferin, TSL.

INTRODUCTION

If one sees luminous sparkles in the wake of a boat or in splashing waves on the beach, they are mostly from the dinoflagellates, an important component of marine phytoplankton. Indeed, these dinoflagellates are responsible for most of the bioluminescence observed on the ocean's surface and are known to occur globally. They are very abundant during red tides and are believed to use their light as a burglar alarm to attract predators to animals grazing on them. Though these dinoflagellates play an important role in the food chain of marine ecosystems, the ecological importance of their bioluminescence is highlighted in this chapter.

The dinoflagellates or fire algae are members of phytoplankton and are generally yellow-brown in color. Typically these organisms contain some photosynthetic and possess two different flagella, which are ribbon-shaped. Of about 2000 known species of dinoflagellates, 1,555 (90%) free-living species are found in the seas and oceans. These dinoflagellates are mostly microscopic ranging from 15 to 40 microns. The largest species *Noctiluca scintillans* (sea sparkle, sea ghost, or fire of the sea), may have a size of about 2 mm in diameter. Several species of marine dinoflagellates have been reported to cause harmful algal blooms (HABs) like anoxia red tides, believed to cause direct effects on fish in oceans worldwide, by damaging their gills or by promoting low dissolved-oxygen concentrations. Further, these red tides may also negatively effect the economy and human health. Seafood poisoning in humans is invariably caused by consuming toxin-containing seafood contaminated with marine dinoflagellates. According to the species of toxigenic dinoflagellates, the poisoning syndromes are named as paralytic (PSP), diarrhetic (DSP), neurotoxic (NSP), azaspiracid shellfish poisoning (AZP) and ciguatera fish poisoning (CFP). Besides these well-known poisoning types, new dinoflagellate toxins, such as yessotoxin (YTX) and palytoxin (PTX), have also been reported from certain species of marine dinoflagellates [14, 37].

BIOLUMINESCENCE IN DINOFLAGELLATES

One of the fascinating characteristics of the marine dinoflagellate is that they can produce and emit blue light during night hours in response to mechanical stimulation, which may even be the turbulence produced by a small air bubble popping. The intracellular bioluminescence of these dinoflagellates is due to its tetrapyrrole luciferin, which is enzymatically oxidized in the presence of dinoflagellate luciferase [38]. Bioluminescence has been reported in 81 species (Table **1**) of 18 dinoflagellate genera [13] and the emission of light is very common among the species of *Alexandrium* such as *Alexandrium affine, Alexandrium acatenella, Alexandrium catenella, Alexandrium fundyense, Alexandrium tamarense, Alexandrium monillatum, Alexandrium ostenfeldii* and *Alexandrium fraterculus*. In these dinoflagellate species, the capacity for bioluminescence is said to remain almost constant throughout the night and is completely restored in 1 hour after exhaustive stimulation [39]. The values of total stimulable luminescence in these dinoflagellates have been reported to range from 10^7 to 10^{10} photons per cell [40]. These dinoflagellates are very sensitive to motion induced by ships or fish-like organisms, and they are said to respond with rapid brilliant flashes, and the glow (Fig. **1**) thus caused is sometimes seen in the wake of a ship.

Table 1. Bioluminescent marine dinoflagellates [13].

Class	Order	Family	Species
Dinophyceae	Prorocentrales	Prorocentraceae	*Prorocentrum micans*
-	Gymnodiniales	Gymnodiniaceae	*Gymnodinium flavum*, G.sanguineum
-	-	Polykrikaceae	*Polykrikos kofoidii,, P. schwartzii*
-	Noctilucales	Noctilucaceae	*Noctiluca scintillans`*
-	Pyrocystales	Pyrocystaceae	*Dissodinium pseudolunula, Pyrocystis acuta, P.lunula,P.fusiformis,P.noctiluca*
-	Peridiniales	Pyrophacaceae	*Fragilidium heterolobum, Pyrophacus horologium*
-	-	Ceratocorythaceae	*Ceratocorys horrida*
-	-	Ceratiaceae	*Ceratium furca, C.candelabrum,C.furca,C.fusus,C.gibberum,C.horridum, C.lunula, C.tripos*

(Table 1) cont.....

Class	Order	Family	Species
-	-	Goniodomataceae	*Alexandrium acatenella, A.catenella, A.fraterculus, A.monilatum, A.ostenfeldii, A.tamarense, Pyrodinium bahamense, Triadinium polyedricum*
-	-	Gonyaulacaceae	*Gonyaulax digitali, G.excavta, G. grindleyi, G.hyalina, G. monacantha, G.monilata, G.parva, G.polygramma, G. scrippsae, G. sphaeroida, G.spinifera, Lingulodinium polyedrum, Peridiniella catenate*
-	-	Peridiniaceae	*Glenodinium sp., Protoperidinium antarcticum, P,bipes, P.brevipes, P.pyrifore, P.brochii, P.creasus, P.claudicans, P.conicoides, P.conicus, P.curtipes, P.crassipes, P.depressum, P.divergens, P.elegans, P.eugrammum, P.excentricum, P.exiquipes, P.globulum, P.granii, P. heteracanthus, P.huberi, P. leonis, P.minutum, P.nudus, P.oceanicum, P.ovatum, P.pacificum, P.pallidum, P.pellucidum, P.pentagonum, P. punctulatum, P.pyriforme, P.saltans, P.seta, P. punctulatum, P.sournia, P. steinii, P. subinerme, P. thulesense, P.tubum*

Fig. (1). Bioluminescence due to dinoflagellates. Image credit: Josh Myers, Florida Adventure (Reproduced with permission).

Emission Maxima and Total Stimulable Luminescence in Dinoflagellates: The emission maxima (λmax) and Total Stimulable Luminescence (photons/cell s^{-1}) observed for the different species of luminescent dinoflagellates have been reported to vary from 472 to 480 nm and from 2.8 x 10^8 to 2.5 x 10^{11} photons/cell s^{-1} respectively (Tables **2** and **3**).

Table 2. Emission maxima of marine dinoflagellates.

Species	Emission maxima (λmax)	Ref
Dissodinium pseudolunula	472 nm	[13]
Dissodinium sp.	474 nm	[41]
Lingulodinium polyedrum	480 nm; 472-474 nm	[13]
-	472nm, 474nm	[42]
-	479nm	[10]
Noctiluca scintillans	470nm	[10]
Peridiniella catenata	480 nm	[13]
Protoperidinium depressum	480 nm	[13]
Protoperidinium divergens	480 nm	[13]
Protoperidinium ovatum	480 nm	[13]
Protoperidinium pallidum	480 nm	[13]
Protoperidinium steinii	480 nm	[13]
Polykrikos schwartzii	480 nm	[13]
Pyrocystis acuta	474 nm	[13]
-	473-475nm	[10]
-	474 nm	[41]
Pyrocystis fusiformis	471nm	[42]
-	471-472 nm	[13]
-	473-475nm	[10]
-	474 nm	[41]
Pyrocystis lunula	472nm	[42]
-	479nm	[10]
Pyrocystis noctiluca	472nm	[42]
-	475,480,473-475nm	[10]
-	473-474nm	[13]
-	474 nm	[41]
Pyrodinium bahamense	479nm	[10]

Table 3. Total Stimulable Luminescence (TSL) in marine dinoflagellates [43].

Species	TSL (photons/cell s^{-1})
Ceratium fusus	5.3×10^8
Ceratium horridum	5.3×10^8
Gonyaulax polygramma	1.6×10^9
Gonyaulax monacantha	6.6×10^8
Gonyaulax scrippsae	3.6×10^8
Gonyaulax sp.	6.6×10^8
Noctiluca scintillans	2.5×10^{11}
Noctiluca sp.	1.3×10^{10}
Protoperidinium globulus	1.1×10^9
Protoperidinium oceanicum	7.1×10^9
Protoperidinium ovatum	1.6×10^9
Protoperidinium palladium	2.8×10^8
Protoperidinium steinii	1.1×10^9
Protoperidinium crassipes	1.9×10^9
Protoperidinium leonis	1.4×10^9

LUMINESCENT MARINE DINOFLAGELLATES

Family: Prorocentraceae

Prorocentrum micans

This luminescent species (Fig. **2**) is found in neritic and estuarine waters and in oceanic environments. It is a cosmopolitan species distributed in cold temperate to tropical waters. It is a bivalvate species and is highly variable in shape and size. Cells are more or less heart shaped; rounded anteriorly and pointed posteriorly; and broadest around the middle . It is strongly flattened with a well developed winged apical spine. Cells are medium-sized (max. 70 µm long and 50 µm wide). Its cell surface is rugose, covered with shallow minute depressions. Numerous tubular trichocyst pores are also present in short rows arranged radially.

Fig. (2). *Prorocentrum micans.* Family: Gymnodiniaceae.

Gymnodinium flavum

This coastal water species is found in the North America, South America and Australia.. Its rounded body has a wide, almost straight girdle and a short sulcus. Surface layer of the body is more or less hardened. There are many and ovoidal Chromatophores. This luminescent species has a maximum length of 40 um and breadth of 40 um.

Gymnodinium sanguineum

This luminescent species (Fig. **3**) is cosmopolitan and is commonly found distributed in temperate to tropical neritic waters. It is an athecate species (*i.e.* without thecal plates) and is highly variable in size and shape. Cells are large, fairly dorso-ventrally flattened and is roughly pentagonal. An apical groove is present. Epitheca and hypotheca are nearly equal in size. Epitheca is rounded and conical, and hypotheca is deeply indented by the sulcus creating two posterior lobes. Cells range in size from 40-80 μm in length.

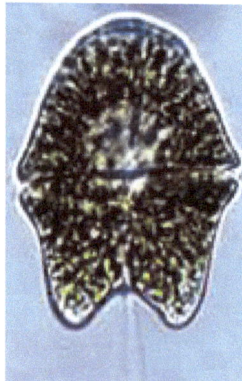

Fig. (3). *Gymnodinium sanguineum.* Family: Polykrikaceae.

Polykrikos schwartzii

This species (Fig. **4**) forms pseudocolony with 2-8 zooids with 1-4 nuclei. Individual zooids are not conspicuous. Cells contain nematocysts-taeniocysts. Its length may range from 100 μm to 150 μm. It has a resting spore in its life cycle with a length of 80-160 μm and width of 50-80 μm.

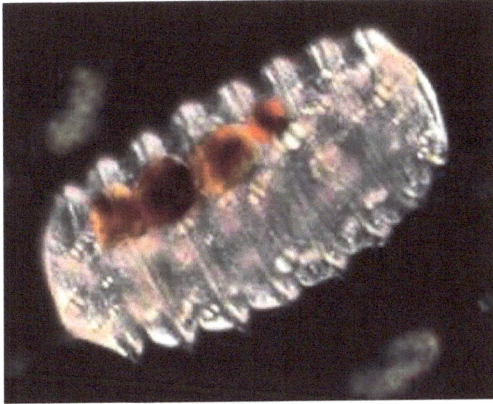

Fig. (4). *Polykrikos schwartzii.* Family: Noctilucaceae. Image courtesy: Malin Mohlin (Reproduced with permission).

Noctiluca scintillans (= Noctiluca miliaris)

This species which is also known as "sea sparkle," "sea fire," "sea ghost" (Fig. **5**) has worldwide distribution. Cells of this species are solitary with striated tentacles and have a diameter ranging from 200 to 2000 μm. It is often found along the coast, in estuary, and shallow areas of the continental shelf. It is a heterotrophic species feeding on algae, plankton and bacteria. It is known for its bioluminescence (Fig. **6**) which is used as a defense mechanism. The bioluminescent characteristic of this species is produced by a luciferin-luciferase system located in thousands of spherically shaped organelles, or "microsources", which are located throughout its cytoplasm. Though the blooms of this species may cause red tide, all blooms associated with this species are not red. The color of this species is partly derived from the pigments of its vacuole. For example, green tides of this species result from the populations containing green-pigmented prasinophytes (green algae, Subphylum Chlorophyta) that are living in their vacuoles. Though this species is not toxic, toxic blooms have been linked to massive fish and marine invertebrate kills. This is largely due to the release of its accumulated ammonia in the surrounding water. Extensive toxic blooms of this species have been reported off the east and west coasts of India, where large scale decline of fisheries has been implicated.

Fig. (5). *Noctiluca scintillans. Image credit: Wikipedia.*

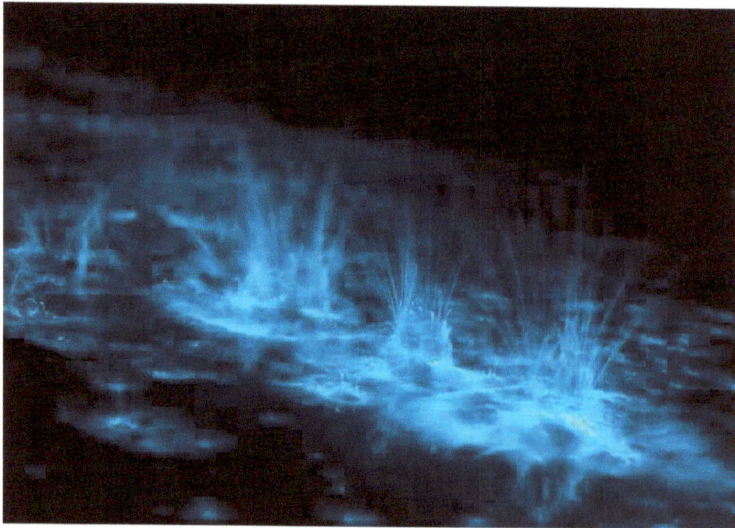

Fig. (6). *Noctiluca scintillans* bioluminescence in action. Image credit: Phil Hart (CC). Family: Pyrocystaceae.

Dissodinium pseudolunula

It is a common luminescent species (Fig. **7**) in the neritic waters of the eastern North Atlantic and the North Sea. It is ectoparasite on the eggs of the marine planktonic copepod Temora longicornis [44]. When the host is absent, the species survives with a distinctive pair of twin resting cysts. It is known to emit light during bloom conditions.

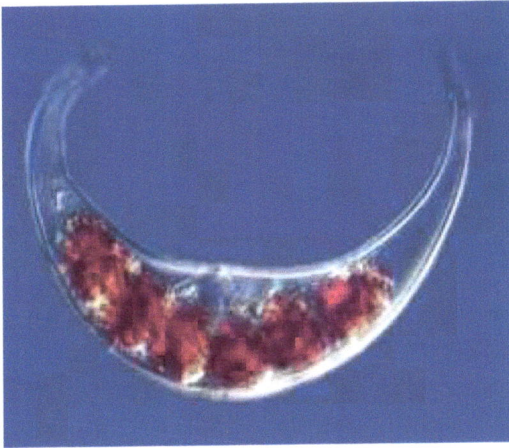

Fig. (7). *Dissodinium pseudolunula.* Image credit: Janina Kownacka (Reproduced with permission).

Pyrocystis acuta

Cells of this species have a fusiform shape. Theca of this species closely resembles that of *P. fusiformis*. The asexual life cycle of this species showed a relatively simple alternation of a transient motile thecate cell-stage with a larger fusiform vegetative cell stage.

Pyrocystis fusiformis

Cells of this luminescent species (Fig. **8**) are tapered or spindle-shaped. It is a non-motile, a characteristic feature of all members of the family Pyrocystaceae. These organisms lose their flagellum by the time they become adults. It is approximately 970 x 163 µm long with a spherical diameter of 374 µm. Interestingly the cell's chloroplasts change the cell's shape as they move closer to the cell's wall in the daytime and retract towards the nucleus at night. It is autotrophic getting its energy from the sun through photosynthesis. It emits bright blue light [45].

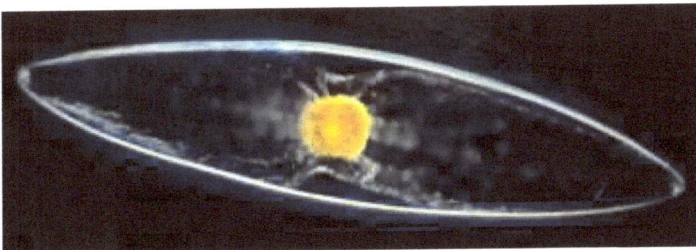

Fig. (8). *Pyrocystis fusiformis.* Image credit: Wikipedia.

Bioluminescence: This species is known for its spontaneous and stimulated bioluminescence. It is reported that in the bioluminescence of this species, the number of flashes per cell was 23-62; its flash duration (ms) as 210 ; and maximum flash intensity as 690 x 109 photons s^{-1} [46].

Pyrocystis lunula

It is a medium-sized luminescent species (Fig. **9**) of *Pyrocystis* with a size ranging between 77 and 280 μm. The shape of the horns is blunt and not acute as in the other species. It is characterized by a normal asexual reproduction with alternations of coccoid cells and morphologically different transitory reproductive stages. In this species, the reproductive bodies are athecate-uniflagellate planospores. After full stimulation during the dark period, a single cell of this species has been reported to emit approximately 4×10^9 photons, which is greater than the light emitted by Pyrodinium bahamense and Lingulodinium polyhedra [47]. This species is considered a model organism due to its bioluminescence capacity linked to circadian rhythms.

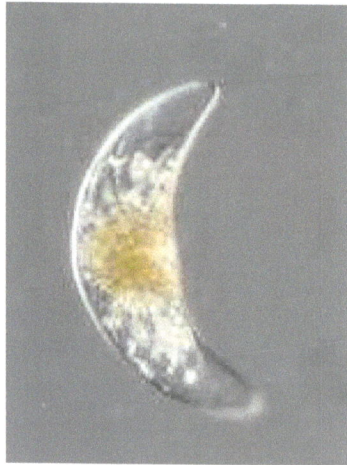

Fig. (9). *Pyrocystis lunula.* Image courtesy: Hanna Mossfeldt, Nordic Microalgae.

Pyrocysti pseudonoctiluca (= Pyrocystis noctiluca)

Image not available

This bioluminescent species has been reported to produce 23-62 light flashes per second and this light is used by this species as a defense mechanism. The original description has not been documented

Family: Pyrophacaceae

Fragilidium heterolobum

Image not available

This luminescent species is a common species in the Californian waters of the Pacific. It is a medium-sized species with a very high number of cingular plates (twelve). The generic name refers to the suddenness with which it is known to shed its plates.

Pyrophacus horologium

This luminescent species (Fig. **10**) which has a diameter of about 80 um is commonly occurring on the Mediterranean coast of Spain. The cell of this species is discoidal and lens-shaped. Both epitheca and hypotheca are equal. In this species, thecal plate numbers are fewer than the other species of this genus. Only a single posterior antapical plate is usually seen.

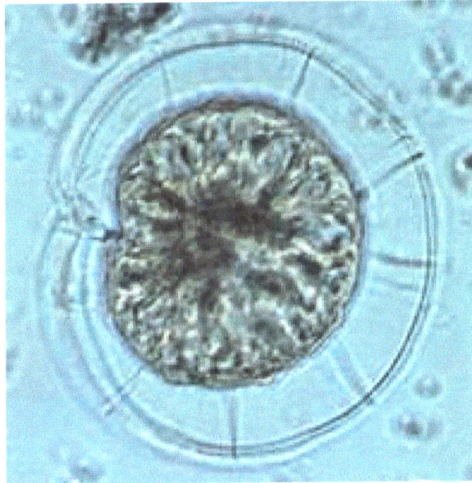

Fig. (10). *Pyrophacus horologium.* Family: Ceratocorythaceae. Image credit: Flickr.

Ceratocorys horrida

This luminescent species (Fig. **11**) is widespread in temperate to tropical ocean waters in the neritic zone. The body of this thecate species may be round in shape, or angular. It has 6 apical spines originating from its body which may vary in length. Its cell length and width is 64 um and 53 um respectively and the corresponding diameter is 70 um.

Fig. (11). *Ceratocorys horrida*. Image credit: Wikipedia.

Bioluminescence: This species is known for its spontaneous and stimulated bioluminescence. It is reported that in the bioluminescence of this species, the number of flashes per cell was 7; its flash duration (ms) as 184; and maximum flash intensity as 9.2 x 109 photons s^{-1} [46].

Family: Ceratiaceae

Tripos furca (= *Ceratium furca*)

This luminescent species (Fig. **12**) has a worldwide distribution. The body of the cell of this species is straight and its epitheca is tapering gradually into an anterior horn. It is found as solitary cells or in pairs. Its length and width may range from 70 to 200 µm and from 30 to 50 µm respectively. It is a bioluminescent specie

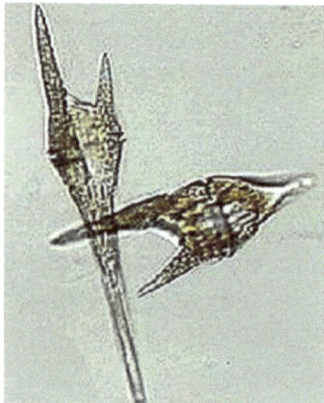

Fig. (12). *Tripos furca*. Image courtesy: Mats Kuylenstierna, Nordic Microalgae.

Tripos fusus (= *Ceratium fusus*)

It is a long fusiform celled luminescent species (Fig. **13**) with a worldwide distribution. Body shape of this species largely resembles that of the Ceratium fusus . The difference however, is that its epitheca is bulbous in shape and extends to a strong apical horn cone. The apical horn is fairly short compared to the antapical horn. Its length and width may range from 200 to 540 µm and from 5 to 40 µm respectively.

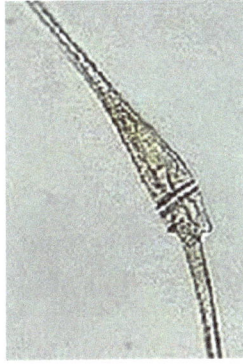

Fig. (13). *Tripos fusus*. Image courtesy: Mats Kuylenstierna, Nordic Microalgae.

Bioluminescence: This species is known for its spontaneous and stimulated bioluminescence. It is reported that in the bioluminescence of this species, the number of flashes per cell was 2; it is flash duration (ms) as 239 ; and maximum flash intensity as 1.1 x 109 photons s^{-1} [46].

Tripos horridus (= *Ceratium horridum*)

It is a solitary, luminescent species (Fig. **14**) and it has worldwide distribution. The cell of this bioluminescent species has open-ended horns and an almost straight apical horn. Its length and width may range from 280 to 360 µm and from 40 to 50 µm respectively

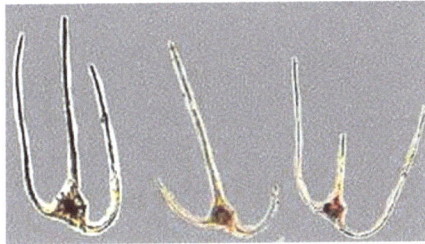

Fig. (14). *Tripos horridus*. Image courtesy: Mats Kuylenstierna, Nordic Microalgae. Family: Goniodomataceae.

Alexandrium monilatum

Image not available

It is a very distinctive chain-forming species occurring in long chains of 2-80 cells. Single cells which are in the size between 28-52 µm in length and 33-60 µm in transdiameter width are flattened anterio-posteriorly. Thecal plates are thin with many delicate pores. It is a common HAB (harmful algal bloom) species of the southern Atlantic and Gulf coasts of the U.S. A dense bloom of this species has also caused magnificent bioluminescence "glowing waters" in the waters of lower Chesapeake Bay during August 2015 [48].

Alexandrium minutum

Cells of this luminescent species (Fig. **15**) are small, somewhat dorsoventrally flattened, nearly spherical to ellipsoidal, and are rarely longer than wide. Cells are single with thin thecal plates . Thecal surface ornamentation which is mostly confined to the hypotheca can vary from light to heavy reticulation with small scattered pores. Intercalary bands are seen. The size of the cells ranges between 15-30 µm in length and 13-24 µm in transdiameter width. This species has been reported to cause discoloration of Samil Beach waters of Vigo, Spain (red tide) during day time and bioluminescent during the night with cell densities ranging from 20 to 120 x10^6 cells L^{-1} [49].

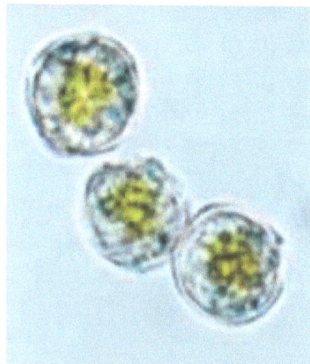

Fig. (15). *Alexandrium minutum.*

Alexandrium ostenfeldi

This luminescent species (Fig. **16**) is nearly spherical. Though these cells are single, they are often found in two-celled colonies. Both the epitheca and hypotheca are equal in height. It has thin thecal plates. Cells range in size between 40-56 µm in length and 40-50 µm in transdiameter width. It is a toxic, harmful bloom forming species capable of emitting light.

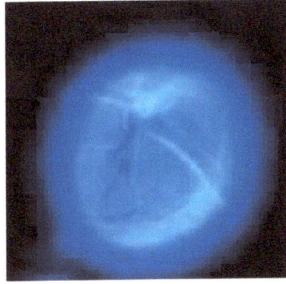

Fig. (16). *Alexandrium ostenfeldi.* Image credit: Janina Kownacka (Reproduced with permission).

Pyrodinium bahamense

This chain-forming luminescent species (Fig. **17**) is mainly distributed in tropical areas of both hemispheres. Cells are polyhedral and irregularly rounded, with strong crests along the sutures. Most cells have a well-developed left antapical spine and a smaller right spine which is an extension of the sulcal list. Its theca has a granular surface and numerous trichocyst pores. This species has been reported to bloom commonly during the rainy summer in shallow with coastal underground drainage. In its life cycle, both motile and cyst stages are seen. This species is also characterized by a high bioluminescence

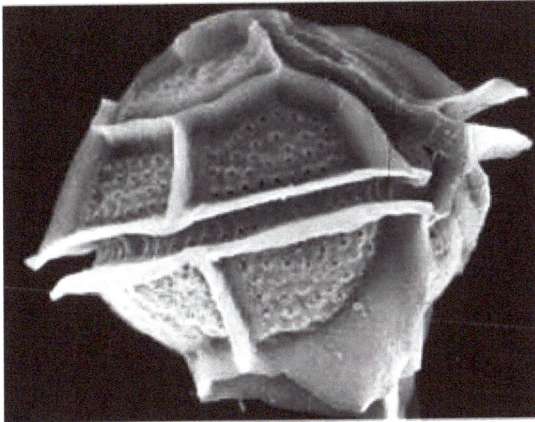

Fig. (17). Pyrodinium bahamense. Image credit: Wikipedia. Family: Gonyaulacaceae.

Gonyaulax polygramma

Image not available

Cells of this luminescent species are elongate and pentagonal; and are medium-sized. Its tapered epitheca which bears a prominent apical horn, exceeds the symmetrical hypotheca. While longitudinal ridges ornament the thecal surface, reticulations are present between the ridges . In the old cells, longitudinal ridges may be thick and spinulous. Cells of this species range in size from 29-66 μm in length and 26-56 μm in dorsoventral depth [50]. It is a red tide-forming species. Although it does not produce toxins, its blooms may cause massive fish and shellfish kills due to anoxic (lack of oxygen) conditions. It is known to emit light during dark hours in the nearshore waters.

Gonyaulax scrippsae

This luminescent species (Fig. **18**) has worldwide distribution. Cells are solitary and possess longitudinal striation and vertical bars across the girdle. Antapical spines are absent and the cyst is present in its life cycle. The length and width of the cells have a size range of 20-57 μm and 27-40 μm respectively.

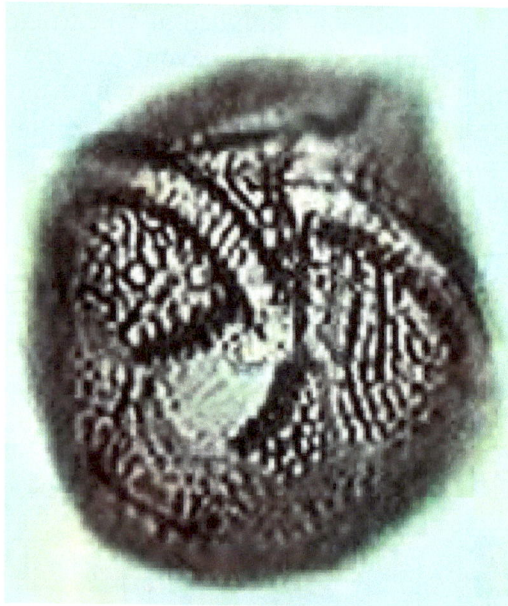

Fig. (18). *Gonyaulax scrippsae*. Image courtesy: Mats Kuylenstierna, Nordic Microalgae.

Gonyaulax spinifera

Cells of this luminescent species (Fig. **19**) are solitary and their length and width ranges from 24-50 μm and 30-40 respectivley. This toxic species which is commonly occurring in the Adriatic Sea and New Zealand waters has been recently found associated with YTX production to the tune of 33.4 pg cell−1 . Its taxonomy remains largely unresolved [51].

Fig. (19). *Gonyaulax spinifera.* Image courtesy: Mats Kuylenstierna, Nordic Microalgae.

Lingulodinium polyedrum (= *Gonyaulax polyedra*)

Cells of this luminescent species (Fig. **20**) are angular and polyhedral with ridges along sutures. This motile photosynthetic species often causes red tides in southern California, leading to bioluminescent displays on local beaches at night. These tiny organisms emit light when stressed say by the lapping of waves, the carving action of a surfboard or other. This species has been reported as a model organism for studying clocks in single cells [52]. This species has also been responsible for the production of Yessotoxins (YTXs), a polyether toxin. This toxin normally gets accumulated in shellfish causing Paralytic Shellfish Poisoning (PSP).

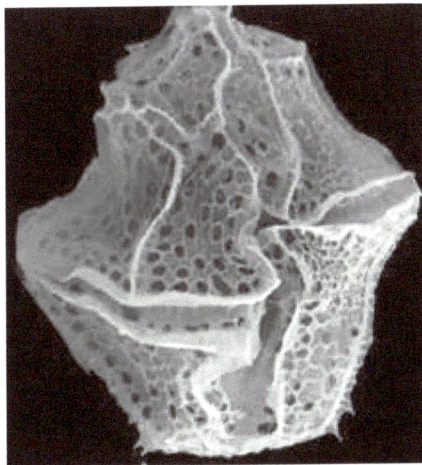

Fig. (20). *Lingulodinium polyedrum.* Image credit: Morseds, Wikipedia.

Bioluminescence: This species is known for its spontaneous and stimulated bioluminescence. The observed bioluminescence of this species is shown in Fig. (21). It is reported that in the bioluminescence of this species, the number of flashes per cell was 2-3 ; its flash duration (ms) as 100-150 ; and maximum flash intensity as 0.19 x 109 photons s^{-1} [46].

Fig. (21). Bioluminescence of *Lingulodinium polyedra*. Image courtesy: Photo by NN, Nordic Microalgae CC.

Peridiniella catenata

This luminescent species (Fig. **22** and **23**) has a wide distribution in cold and temperate waters. It appears as solitary or in chains. Its length and width may range from 20 to 35 µm and 20 to 35 µm respectively.

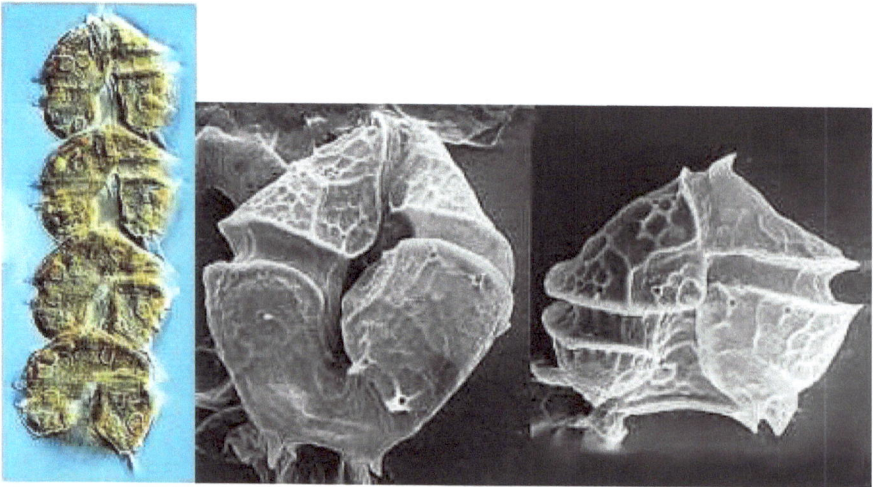

Fig. (22). *Peridiniella catenata*. Image courtesy: Mats Kuylenstierna, Nordic Microalgae.

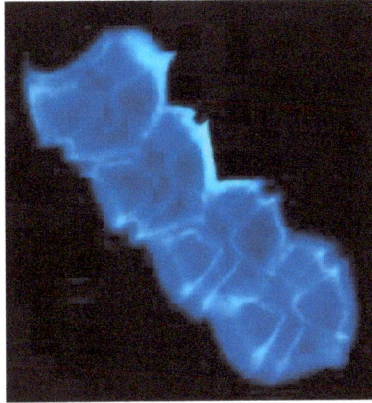

Fig. (23). *Peridiniella catenata* Luminescent form. Image credit: Janina Kownacka (Reproduced with permission).

Protoceratium reticulatum (= *Gonyaulax grindleyi*)

This luminescent species (Fig. **24**) is widely distributed in temperate and tropical areas; sub-boreal waters; and also in North and South Atlantic, Pacific and Indian Oceans and in their nearby seas. Cells of this solitary species are oval to polygonal. Both the hypotheca and epitheca are oval to bowl-shaped. Cell surface is highly reticulated and cell has numerous brown chloroplasts. It is a heterotrophic dinoflagellate with asexual reproduction. Its size is varying from 28 to 43 μm in length and from 25 to 35 μm in width. This species has been reported to produce a disulfated polyether called yessotoxin (YTX), which is harmful to human health, aquaculture development and coastal environments. This bloom is believed to be initiated mainly by a decrease in salinity caused by rainfall.

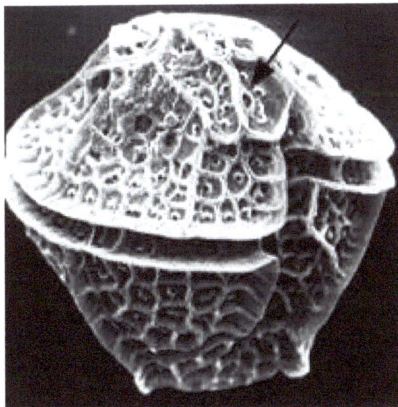

Fig. (24). *Protoceratium reticulatum*. Image courtesy: Mats Kuylenstierna, Nordic Microalgae. Family: Peridiniaceae.

Protoperidinium crassipes

This luminescent species (Fig. **25**) is found distributed throughout the world's oceans. Cells are solitary and are with apical horn and antapical horns with spines. The size of the cells ranges from 80 to 100 µm in length and from 80 to 105 µm in width. Though this species does not produce toxins, it is known to contain them because it consumes another toxic dinoflagellate, called *Azadinium*.

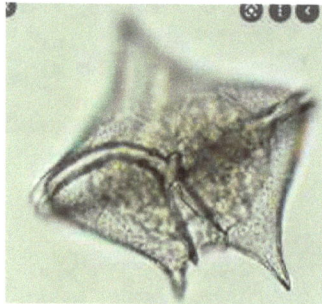

Fig. (25). *Protoperidinium crassipes*. Image courtesy: Flickr.

Protoperidinium curtipes

This luminescent species (Fig. **26**) has worldwide distribution. Theca of this species is compressed and angular at the plane of its cingulum. Epitheca forms an apical horn. The cingulum is slightly left-handed and excavated. Hypotheca forms two conical antapical horns with robust spines at their tips. The left horn is somewhat shorter than the right one. The sulcus is deep and forms a strong indentation between the horns. There is one more spine-like protrusion between the antapical protrusion of the left sulcal list and the spine of the left antapical horn. Ornamentation of the plates is reticulated. Theca length and width range from 95 to 107 µm and from 77 to 96 µm respectively.

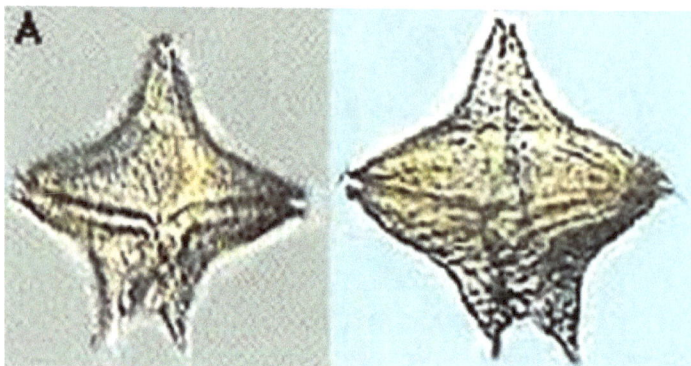

Fig. (26). *Protoperidinium curtipes*. Image courtesy: Mats Kuylenstierna, Nordic Microalgae.

Protoperidinium depressum

This solitary, luminescent species (Fig. **27**) has worldwide distribution. Theca is compressed and rounded at the plane of the cingulum. It also has a strong ventral indentation which t gives theca a bilobial form. Epitheca forms a robust apical horn. Cingulum is narrow and lefthanded with wide lists. While the left arc of the cingulum has an apical direction, the right arc has an antapical direction. Hypotheca forms two robust, conical, pointed, diverging antapical horns. The left antapical horn is slightly shorter than the right one. Sulcus is deep and forms a strong indentation of the antapical margin of the hypotheca, between the horns. Left sulcal list is well developed. Ornamentation of the plates is reticulated. Theca has a length of 129-149 μm and a width of 100-129 μm. It is a luminescent species.

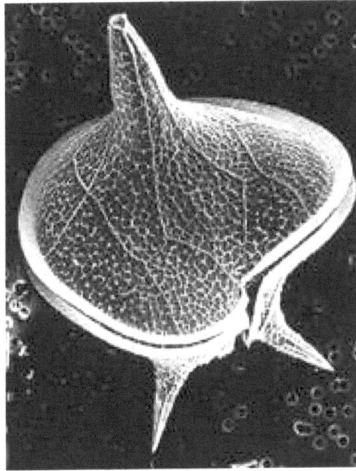

Fig. (27). *Protoperidinium depressum*. Image courtesy: Mats Kuylenstierna, Nordic Microalgae.

Protoperidinium divergens

This solitary, luminescent species (Fig. **28**) has worldwide distribution. Theca is angular centrally. Epitheca forms an apical horn. Cingulum is circular and excavated. Hypotheca forms two conical and diverging antapical horns. Sulcus is deep and forms a strong indentation of the antapical margin, between the horns. Ornamentation of the theca is reticulated. Theca has a length of 75-95 mm and width of 50-65 mm. It is a luminescent species.

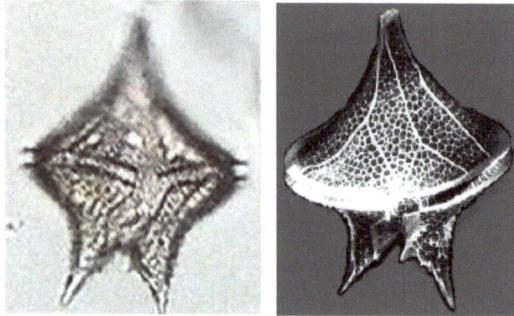

Fig. (28). *Protoperidinium divergens*. Image courtesy: Mats Kuylenstierna, Nordic Microalgae.

Protoperidinium leonis

This solitary, luminescent species (Fig. **29**) species has worldwide distribution and it has a resting spore in its life cycle. Its length ranges from 50 μm to 95 μm. It is a luminescent species.

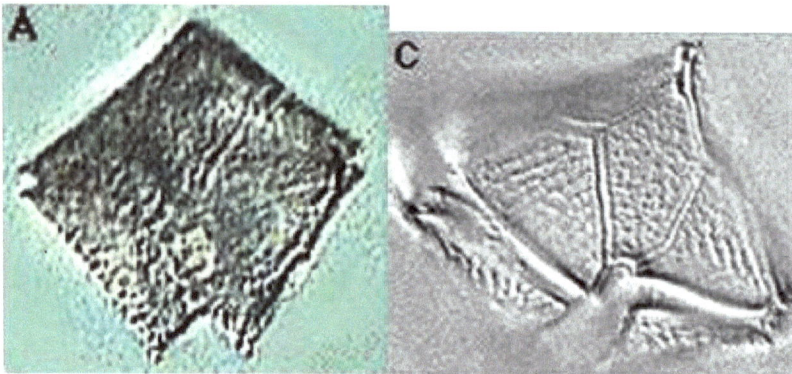

Fig. (29). *Protoperidinium leonis*. Image courtesy: Mats Kuylenstierna, Nordic Microalgae.

Protoperidinium oceanicum

Image not available

This solitary species has worldwide distribution. Theca is centrally rounded. Epitheca forms a long apical horn. Cingulum is narrow and left-handed with wide lists. Hypotheca forms two long, tubular, pointed antapical horns which are diverging. The left horn is slightly shorter and thinner than the right horn. Sulcus is deep and ornamentation of the theca is reticulated with spiny junctions. Theca has a length ranging from 163 to 187 μm and width from 86 to 102 μm. It is a luminescent species.

Protoperidinium ovatum

This solitary, luminescent species (Fig. **30**) has worldwide distribution and it has a size ranging from 54 to 68 μm in length.

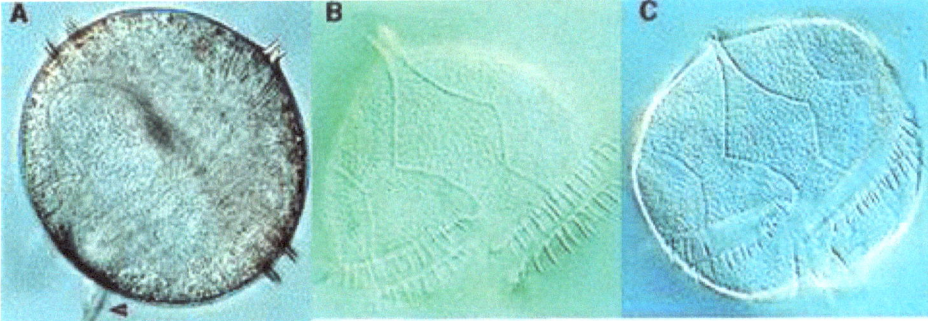

Fig. (30). *Protoperidinium ovatum*. Image courtesy: Mats Kuylenstierna, Nordic Microalgae.

Protoperidinium palladium

This solitary, luminescent species (Fig. **31**) is dorsoventrally compressed and it has worldwide distribution. Epitheca is almost conical with a short apical horn. Cingulum is right-handed with wide lists. Hypotheca is hemispherical with a concave antapical margin and two antapical spines which are solid with wings. Left spine is slightly shorter than the right one. Sulcus is deep and it widens at its lowest part with a well-developed left sulcal list that protrudes antapically. Theca has a length of 97-116 μm and a width of 62-71 μm. It is a luminescent species.

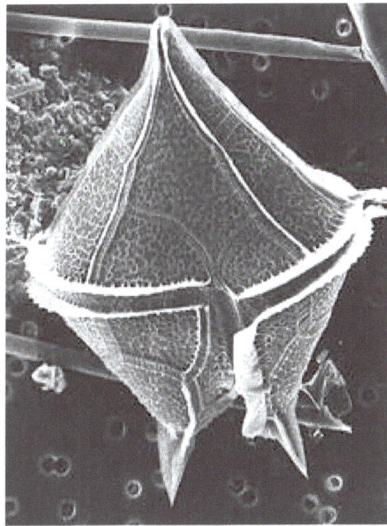

Fig. (31). *Protoperidinium palladium*. Image courtesy: Mats Kuylenstierna, Nordic Microalgae.

Protoperidinium pellucidum

This solitary, luminescent species (Fig. **32**) has worldwide distribution and it has a diameter ranging from 36 to 52 µm.

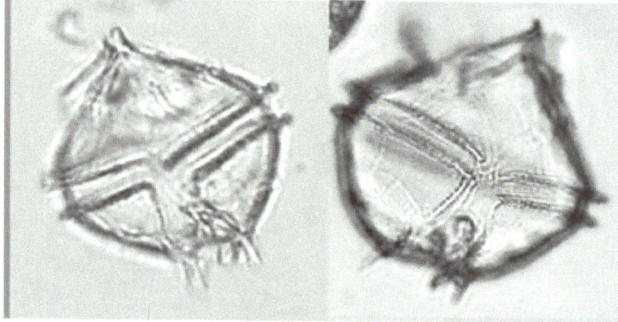

Fig. (32). *Protoperidinium pellucidum.* Image courtesy: Regina Hansen, Nordic Microalgae.

Protoperidinium steinii

This solitary, luminescent species (Fig. **33**) has worldwide distribution. Cells of this species are pyriform with a long apical horn and two three-winged antapical horns. Its length and width range from 39 to 60 µm and 22 to 44 µm respectively

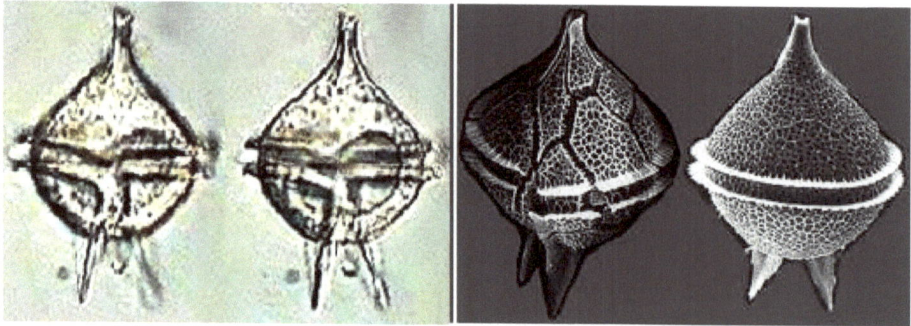

Fig. (33). *Protoperidinium steinii.* Image courtesy: Mats Kuylenstierna, Nordic Microalgae.

CONCLUSION

Though the dinoflagellates have attained their prominence as producers of bioluminescence in the ocean, this fascinating phenomenon has received little attention as per the major findings reported in the last quarter of a century. This has led us to question key paradigms in the field or those that have revealed major gaps in our knowledge relating to dinoflagellate bioluminescence. One such research priority is to understand the structure and origin of luciferin in bioluminescent dinoflagellate species other than P. lunula, This calls for further research in this vital aspect.

CHAPTER 4

Bioluminescent Marine Radiolarians

Abstract: This chapter deals with the identified luminescent species of marine, colonial radiolarians, emission maxima in observed species of marine radiolarians, the description of luminescent marine radiolarian species and their mechanism of bioluminescence.

Keywords: Colonial radiolarians, Emission maxima, Stimulated bioluminescence.

INTRODUCTION

Radiolarians (single-celled marine protists), the members of silica-secreting zooplankton, are like amoebae living in glass houses and are common in shallow oceanic gyres of deep oceans worldwide. However, some species are limited to certain regions and serve as temperature, salinity, and total biological productivity indicators. The skeletal remains of some radiolarians make up a fairly large part of the cover of the ocean floor as siliceous ooze. It is interesting to note that these organisms use calcium-activated photoproteins much like hydromedusae in their bioluminescence. Radiolarians that inhabit great depths in the water column where light is limited or absent typically lack algal symbionts.

Radiolarians (Phylum: Protozoa) are incredibly diverse in their skeletons, ranging from spherical to rod-shaped and radial to bilaterally symmetrical. Their cytoplasmic mass is divided into two regions separated by a perforated membrane. The first of these regions is the central mass or the central capsule, and the second is the extracapsulum, a peripheral layer of cytoplasm surrounding the central capsule. The central capsule contains the organelles such as the mitochondria and vacuoles, while the extracapsulum is characterized by its thread-like extensions of the cytoplasm, *i.e.*, the rhizopodia. It is reported that both the solitary and colonial species of radiolarians can produce light with multiple flashes [38]. A total of 9 taxa of radiolarians (Table **1**) have been reported to possess the characteristic, bioluminescence which is largely due to the mechanical stimuli and their luciferin (coelenterazine) and Ca^{2+} activated photoprotein [27]. Light emission in these organisms is deep blue with peak emissions between 443 and 456 nm wavelengths. Single flashes are 1–2 seconds in duration with species-dependent flash kinetics. Colonies of different species (even families) may display different flash kinetics. The quantal content of single flashes averaged 1×10^9 photons flash-1, and the colonies have shown prolonged light emission. The mean value of bioluminescence potential in these luminescent radiolarians based on total mechanically stimulated bioluminescence has been reported to be 1.2×10^{11} photons colony-1. It is also estimated that colonial radiolaria have the capacity to produce $\approx 2.8 \times 10^{12}$ photons $\cdot$ m^{-2} of sea surface [53].

Table 1. Bioluminescent marine radiolarians [13].

Class	Order	Family	Species
Radiolaria	Phaeogromia	Tuscaroridae	*Tuscaridium cygneum*
-	Spumellarida	Thalassicolidae	*Thalassicola nucleata, Thaiassicolla sp.*
-	-	Thalassothamnidae	*Cytocladus major*
-	-	Sphaerozoidae	*Rhaphidozoum acuferum*
-	-	Collosphaeridae	*Acrosphaera murrayana, Collosphaera huxleyi,* Myxosphaera coerulea, *Siphonosphaera tenera,*
-	Phaeosphaerida	Aulosphaeridae	*Aulosphaera triodon*

Emission Maxima in Luminescent Marine Radiolarians

The emission maxima (λmax) of the different species of luminescent marine radiolarians have been reported to vary from 443 to 458 nm (Table **2**).

Table 2. Emission maxima (λmax) of luminescent marine radiolarians.

Species	λmax (nm)	Ref
Acrosphaera murrayana	443 nm	[23]
Collosphaera huxleyi	456 nm	[23]
Collosphaera spp.	443, 445, 452 nm	[23]
Myxosphaera coerulea	453 nm	[23]
Rhaphidozoum acuferum	458 nm	[23]
Siphonosphaera tenera	450 nm	[23]
Thalassicola nucleata	446 nm	[54]
Thaiassicolla sp.	450nm	[10]

LUMINESCENT MARINE RADIOLARIANS

Acrosphaera murrayana

The shell of this luminescent species (Fig. **1**) is spherical, with large circular or roundish pores of unequal size. Ten to twelve pores are seen in the half meridian of the shell. The margin of every pore is with a coronal of six to nine short and acute spines. No spines are present between the pores. Shell diameter varies from 70 to 190 μm.

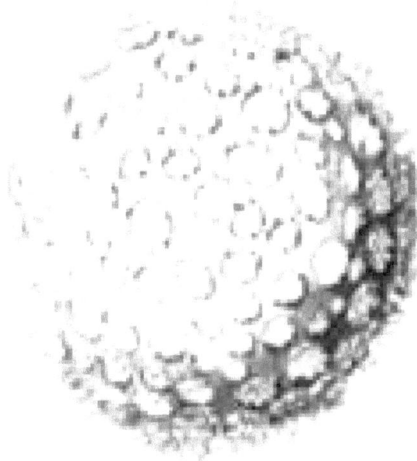

Fig. (1). *Acrosphaera murrayana.* Image credit: Wikipedia

Aulosphaera triodon

The shell of this luminescent species is spheroidal to ellipsoidal with triangular (sometimes square) mesh openings. Its radial tubes are generally smooth and are with 2-4 (rarely 3, seldom 4) straight or slightly curved terminal branches. The diameter of the shell varies from 1.2 to 4.0 mm.

Collosphaera huxleyi

The shells of this luminescent species (Fig. **2**) have small to medium-sized pores that are scattered about the surface only. There are no spines or tubes. The shell diameter varies from 80 to 150 μm.

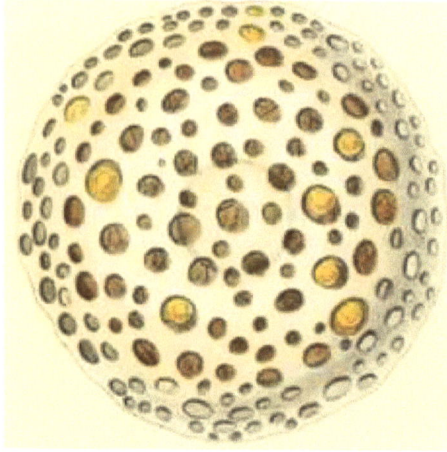

Fig. (2). *Collosphaera huxleyi.* Image credit: Haeckel, Ernst, Wikimedia.

Siphonosphaera tenera

Image NA

In this luminescent species, tubes of variable length are seen on some pores. Tube walls are imperforate. The shell diameter of this species varies from 80 to 120 µm.

Thalassicola nucleata

It is a solitary large spherical cell (Fig. **3**) with a granular appearance. Unlike other species, it lacks both skeleton and spicules. It has a single nucleus at the centre of the intracapsulum containing droplets. Many foamy vesicles are seen in the ectoplasm.

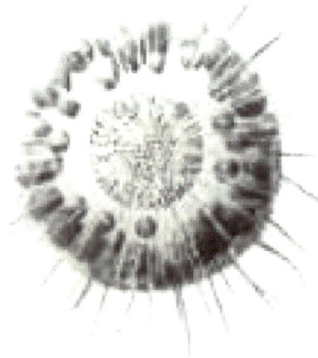

Fig. (3). *Thalassicola nucleata.* Image credit: Camille Dégardin, Muséum national d'histoire naturelle.

Bioluminescence: In this species, the luminescence originates from the outer gelatinous layer [38]. It is reported that its light emission, which is propagated at a rate of 0.5 cm/s^{-1}, originated from the micro sources distributed throughout the extracapsulum. In this species,a single brief mechanical stimulus was found to produce a flash with a mean duration of 5s and a maximum photon flux of ~7 × 108 photons s^{-1}.Total mechanically stimulable luminescence was found to be 5×109 photons per organism [55].

Tuscaridium cygneum

Image NA

This luminescent species with a diameter of 1.2 cm, forms colonies in the deep sea, and it was found to glow when disturbed [56].

CONCLUSION

Though the radiolarians are second only to diatoms as a major source of silicate deposited in the ocean sediments, the planktonic, luminescent radiolarians are so far a neglected component of the marine zooplankton. Variations in the number and kind of luminescent radiolarian species in relation to sediment depth could provide information about climatic and environmental conditions in the overlying water mass. Hence, more detailed studies may be required on this hitherto unknown luminescent species of planktonic radiolarians.

CHAPTER 5

Bioluminescent Marine Cnidarians

Abstract: This chapter deals with the total luminescent fauna of planktonic hydrozoa and scyphozoa, emission maxima in observed species of planktonic hydrozoa and scyphozoa, the description of luminescent marine species of planktonic hydrozoa and scyphozoa and their mechanism of bioluminescence.

Keywords: Emission maxima, Luminescent hydrozoan medusae, Luminescent scyphozoan medusae, Luminescent siphonophores.

INTRODUCTION

Cnidarians which include four main groups *viz.* the almost wholly sessile Anthozoa (sea anemones, corals, sea pens), swimming Scyphozoa (jellyfish), Cubozoa (box jellies), and Hydrozoa (that includes all the freshwater cnidarians), as well as many other marine forms. They were all formerly grouped with ctenophores under the common phylum *viz.* Coelenterata, but their differences in morphology and biology caused them to be placed in a separate phylum *i.e.*, Cnidaria, which was recognized in 2007. Cnidarians are mostly predators, but certain species may also scavenge dead animals or obtain nourishment from intracellular, photosynthetic unicellular algae, called zooxanthellae. Among the cnidarians, several species are venomous, causing human mortalities worldwide. Bioluminescence as a common phenomenon has also been widespread in cnidarians. This chapter deals with the biology and ecology of bioluminescent cnidarians.

All members of the phylum Cnidaria (formerly Coelenterata) are radially symmetrical with a two-layered body made up ofan ectoderm and an endoderm. The three classes of this phylum consist of the Hydrozoa in which the animals have a lifecycle with two phases *viz.* a sessile polyp phase and a free-floating medusoid stage, the class Scyphozoa which includes the medusae thatonly have the medusoid stage in their life cycle, and the class Anthozoa in which the animals are non-planktonic and only have a polyp stage in their life cycle.

Hydromedusae: Generally, the hydromedusae are small, and they are either transparent or lightly pigmented, although some of the deep-sea species are dark red in coloration. Hydromedusae are often common in coastal habitats but they are usually believed to be seasonal. Most coastal hydromedusae are asexually budded off their single-sexed parent hydroids. Each hydroid colony that produces medusae generates only male or female medusae, but not both. The female or male medusae then produce eggs and sperm free-spawned into the sea; the fertilized eggs develop into new hydroids, which are usually benthic (bottom-living). The hydromedusae, therefore, represent only part of the life cycle of each animal. Some open ocean hydromedusae also have hydroids, which may live deep on the sea floor, but some of these oceanic hydroids have found highly specialized substrates to live on, such as little floating clumps of algae, the skin of fishes, or the shells of pelagic snails. Other hydromedusae (typically oceanic or deep water species) do not have a hydroid but have a life cycle in which the fertilized eggs produced by medusae instead develop directly into the next generation of medusae. Such species are sometimes described as "holoplanktonic" carrying out their entire life cycle in the plankton.

Siphonophores: These polymorphic planktonic marine cnidarians inhabit the deep pelagic ocean at depths from ~200 m to the abyssal seafloor (>4,000 m). A typical siphonophore consists of a pneumatophore, nectosome (with nectophores) and siphosome (with its constituents such as bract, gonodendron, tentaculozooid and gastrozooid) . Each loop of the siphosome is called cormidium. Siphonophore colonies have a modular body plan with different zooids performing different functions. Some species of siphonophores have elaboratefluorescent or bioluminescent lures to attract prey.

BIOLUMINESCENT MARINE HYDROZOAN MEDUSAE AND SIPHONOPHORES

A total of about 80 species of hydrozoans are listed below in Table **1**. Both hydrozoan medusae and siphonophores have been reported to be luminescent [13]. In all the planktonic cnidarians, bioluminescence is due to their imidazolopyrazine (coelenterazine) and Ca^{2+-} activated photoprotein. Light production in these organisms is limited to the blue-green wavelengths *i.e.* from 440 to 506 nm. Further, no significant difference has been reported in the mean wavelength between scyphomedusae (474.0 nm) and hydromedusae (473.7 nm). All the luminous hydrozoans possess a photoprotein system of light production (where the luciferin is coelenterazine) *e.g.* obelin in *Obelia geniculata* and clytin in *Clytia hemisphaerica*. As all bioluminescent hydrozoans are believed to possess green fluorescent protein (GFP) in their photocytes, their bioluminescent flash is green [27]. The mean emission max for all the observed species of hydromedusa was found to be 473.8 nm (range 443 - 505 nm). Among the hydromedusae, the four shortest-wavelength species were found to be the Trachymedusae of the family Halicreidae. On the other hand, the longest-wavelength medusae are Leptomedusae such as the common and well-known, *Aequorea forskalea* and *Phialidium (=Clytia) hemisphaerica,* which bear GFP. Narcomedusae such as *Aegina citrea* and *Solmissus* spp. produced intermediate luminescence spectra, with emission max between 460 and 478 nm [57]. The siphonophore spectra were found distributed bimodally with modes centered at 450.5 nm and at 486 nm (range 442 - 500 nm). The light from deep-dwelling siphonophore species showed significantly shorter wavelengths than light from shallow species as in the case of the medusae [57]. Among the siphonophores, the shortest wavelengths have been recorded from the species of *Apolemia* and *Bargmannia,* whereas the light with the longest wavelength was from the epipelagic *Muggiaea* sp. Further, in the siphonophores, there was no significant relation between luminescence and species. For example calycophorans and physonects are more or less equally well represented by species of short and long wavelengths . Three species of siphonophores, *Abylopsis tetragona, Bargmannia elongata,* and *Frillagalma vityazi,* were observed to produce multiple colors of luminescence. In *Abylopsis tetragona,* a major peak with 489 nm, and a secondary peak at 450 nm have been recorded. *Vogtia glabra* produced a unimodal emission with an emission max near 450 nm. It is worth noting that the posterior end of some species of siphonophores produced predominantly green light while the anterior end produced more blue light [57].

Table 1. Bioluminescent marine cnidarians: Hydrozoan medusae and siphonophores [13]

Class	Order	Family	Species
Hydrozoa	Hydroida	Tubulariidae	*Euphysora valdiviae*
-	-	Bougainvilliidae	*Bougainvillia carolinensis*
-	-	Calycopsida	*Bythotiara depressa*
-	-	Pandeidae	*Leuckartiara octona, Pandea conica*
-	-	Mitrocomidae	*Cosmetira pilosella, Halistaura cellularia, Halopsis ocellata, Mitrocoma cellularia, Mitrocomella polydiademata*
-	-	Campanulariida	*Obelia lucifera, Phialidium (=Clytia) hemisphaerica, Phialidium gregarium*
-	-	Aequoreidae	*Aequorea forskalea, Aequorea macrodactyla, Aequorea victoria, Aequorea vitrina*
-	-	Phialuciidae	*Octophialucium funerarium*
-	-	Eutimidae	*Eutonina indicans, Tima bairdi, Tima saghalinensis*

(Table 1) cont.....

Class	Order	Family	Species
-	Trachylina	Geryonidae	*Halitrephes maasi, Halitrephes valdiviae, Liriope tetraphylla*
-	-	Rhopalonematidae	*Colobonema sericeum, Crossota alba*
-	-	Halicreatidae	*Halicreas minimum, Haliscera conica*
-	-	Cuninidae	*Cunina globosa, Solmissus albescens, Solmissus incisa, Solmissus marshalli*
-	-	Aeginidae	*Aegina citrea, Aeginura grimaldii, Solmundella bitentaculata*
-	-	Solmarisidae	*Pegantha clara, Pegantha laevis, Solmaris leucostyla*
-	Siphonophora	Rhizophysidae	*Rhizophysa sp.*
-	-	Apolemidae	*Apolemia sp.*
-	-	Agalmidae	*Agalma okeni, Erenna sp., Frillagalma vityazi, Halistemma amphytridis, Nanomia bijuga, Nanomia cara*
-	-	Pyrostephidae	*Bargmannia sp.*
-	-	Forskaliidae	*Forskalia sp.*
-	-	Prayidae	*Craseoa lathetica, Amphicaryon acaule, Amphicaryon ernesti, Maresearsia praeclara, Nectadamas diomedeae, Nectopyramis natans, Praya dubia, Rosacea plicata*
-	-	Hippopodiidae	*Hippopodius hippopus, Vogtia glabra, Vogtia serrata, Vogtia spinosa*
-	-	Diphyidae	*Chelophyes contorta, Diphyes dispar, Muggiaea sp., Sulculeolaria sp.*
-	-	Clausophyidae	*Clausophyes ovata, Chuniphyes multidentate*
-	-	Abylidae	*Abyla pentagona, Abylopsis eschscholtzii, Abylopsis tetragona, Bassia bassensis,*
-	-	Erennidae	*Erenna sp.*

Emission maxima in luminescent marine hydrozoan medusae and siphonophores:

The emission maxima (λmax) of the different species of marine hydrozoan medusae and siphonophores have been reported to vary from 443 to 550 nm (Table **2**).

Table 2. Emission maxima (λmax) of luminous marine hydrozoan medusae and siphonophores.

Species	λmax(nm)	Ref
Abylopsis eschscholtzii	517 nm	[58]
Abylopsis tetragona	489nm	[57]
Aegina citrea	469nm	[57]
-	459 nm	[23]
Aeginura grimaldii	464nm	[57]

(Table 2) cont.....

Aequorea aequorea	503nm	[57]
-	508nm	[10]
Aequorea macrodactyla	496 nm	[59]
Aequorea victoria	460 nm	[16]
Agalma okeni	550nm	[57]
-	444 nm	[24]
Amphicaryon acaula	487 nm	[13]
Amphicaryon ernesti	487 nm	[13]
Bargmannia sp.	499 nm	[13]
Bougainvillia carolinensis	452 nm	[13]
Bythotiara depressa	488 nm	[57]
Chuniphyes multidentata	481nm	[57]
Colobonema sericeum	494 nm	[60]
Craseoa lathetica	489nm	[57]
Cunina globosus	462nm	[57]
-	494 nm	[60]
Diphyes dispar	464 nm	[13]
Erenna sp.	455nm	[57]
Euphysora valdiviae	464nm	[57]
Frillagalma vityazi	492nm	[57]
Halicreas minimum	469 nm	[57]
Haliscera conica	451nm	[57]
Halistaura cellularia	460nm	[10]
Halistemma amphytridis	451nm	[57]
Halistemma spp.	460nm	[13]
Halitrephes maasi	458nm	[57]
Halitrephes valdiviae	443nm	[57]
Halopsis ocellata	458nm	[57]
Hippopodius hippopus	450nm	[57]
-	455nm	[10]
-	447nm	[23]
Maresearsia praeclara	486nm	[57]
-	473nm	[42]
Mitrocoma cellularia	505nm	[57]
Mitrocomella sp.	500 nm	[13]

(Table 2) cont.....

Muggiaea sp.	500 nm	[13]
Nanomia bijuga	457nm	[57]
Nanomia cara	454 nm	[57]
Nectadamas diomedeae	443 nm	[57]
Nectopyramis natans	447 nm	[57]
Obelia lucifera	509nm	[13]
Octophialucium funerarium	487nm	[57]
Pandea conica	470nm	[57]
Pandea sp.	466 nm	[23]
Pegantha clara	460 nm	[23]
Pegantha laevis	460nm	[57]
Pelagia noetiluea	469 nm	[23]
Phialidium (=Clytia) hemisphaerica	504 nm	[13]
Praya dubia	477nm	[57]
Rosacea plicata (larva)	488 nm	[23]
Solmissus albescens	478nm	[57]
Solmissus incisa	465nm	[57]
Solmissus marshalli	477nm	[57]
Solmundella bitentaculata	477 nm	[13]
Vogtia glabra	448nm	[57]
-	470nm	[10]
Vogtia serrata	451nm	[57]
Vogtia spinosa	470 nm	[13]

Luminescent Marine Hydrozoan Medusae

Aegina citrea

Image credit: NOAA Ocean Exploration & Wikimedia

It is a deep water luminescent species (Fig. **1**) living below 900m and is found distributed in Antarctic, Atlantic, Indo-Pacific and Arctic Oceans. The Bell of this species is pyramidal with four (sometimes 5 or 6) marginal primary tentacles which are inserted into the bell above the margin. Mesoglea is thick at the apex and there are 4 lappets with numerous statocysts. Tentacles and stomach are colorless, whitish, bright yellow, pale pink, mottled brown, or red. The diameter of the umbrella is up to 50mm.

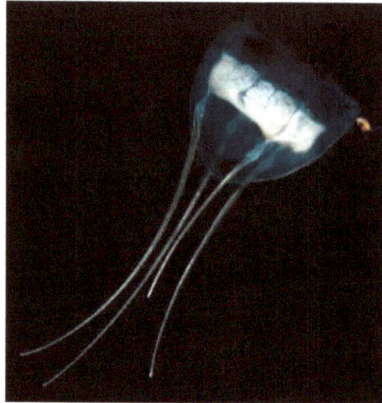

Fig. (1). *Aegina citrea.*

Bioluminescence: It has been reported to produce a series of flashes from one of its four tentacles. These flashes were shorter in duration and were confined initially to the base of the tentacle, until the t-real flash, which propagated distally at about 20 cm s^{-1} [60].

Aeginura grimaldii

Image credit: Jules Richard - Richards, J. Wikimedia

This luminescent species (Fig. **2**) is found distributed in the Atlantic Ocean, Pacific Ocean, and the Arctic Ocean at depths of 660-1200m. Live animals are bright reddish in color, with pale tentacles, and a light red spherical capsule dome which has a dark-colored red body. It measures 4.5cm from the peak of the bell to the end of the tentacle. There are 8 primary tentacles and 3-5 secondary marginal tentacles in each octant of the umbrella margin.

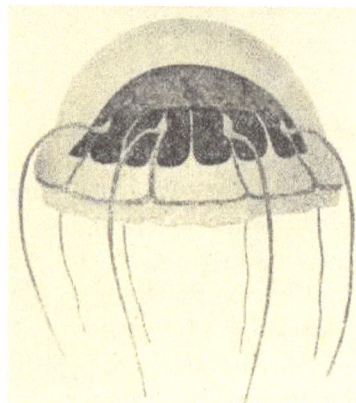

Fig. (2). *Aeginura grimaldii .*

Bioluminescence: When these animals were stimulated with short pulses, they responded with glows lasting 2-6 sec [61]. This luminous hydromedusa has been reported to emit light by intramolecular reaction of the protein named aequorin. This reaction does not involve molecular oxygen and is triggered by the binding of calcium ions. As the photoprotein aequorin is consumed by the reaction, this process cannot be included under the category of enzymatic reaction [4].

Aequorea forskalea (= *Aequorea aequorea*)

Image credit: Leonardo Forbicioni, Wikimedia

This shallow water,1 uminescent species (Fig. **3**) is commonly called the many-ribbed jellyfish. It is found in various temperate and subtropical areas such as the Southwest Atlantic near northern Patagonia; the west coast of Southern Africa; and from the North Sea to the Norwegian Sea.It has a large umbrella which is thick near the center but gradually thin as it reaches the margin of the umbrella. The stomach takes up about half of the overall width of the disc.T his species has 60-160 radial canals. Its gonads run throughout nearly the entire width of this hydrozoan. The number of tentacles is usually fewer than the amount of radial canals per individual. Umbrella may span up to 175 mm across. There are small bulbs scattered around the marginal region while bulbs on the tentacles are conical and elongated.

Fig. (3). *Aequorea forskalea.*

Bioluminescence: Though this species has the bioluminescent protein, aequorin, it is almost colorless. its bioluminescence is largely confined to point sources around the bell margin; and the emissions of all these medusae were blue [60, 62].

Aequorea macrodactyla

Image credit: Gur A. Mizrahi (Reproduced with permission)

This luminescent species (Fig. **4**) is widely distributed in the Indian and Pacific Oceans from Africa to America. It has a biconvex lens umbrella (central disc is lens-shaped) of around 20 mm thick and 6.5 cm to 8.0 cm in diameter. There are 32 straight radial canals and the gonads are linear on both sides of each radial canal. it also has 10–20 broad marginal tentacle bulbs. Marginal tentacles possess cross-shaped bases. The color of the radial canal and the tentacle bulbs are milky colored and other parts are mostly transparent [63].

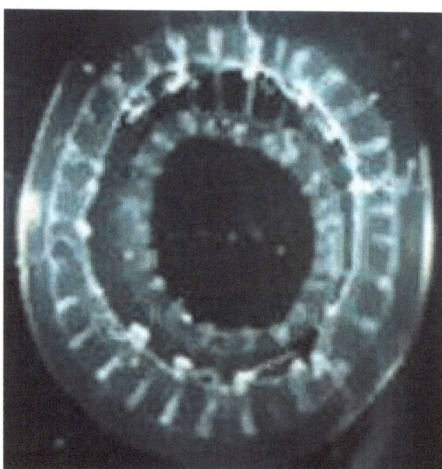

Fig. (4). *Aequorea macrodactyla .*

Bioluminescence: The luminescent system of this species consists of a green fluorescent protein (GFP) and one or more aequorins. The purified photoprotein of this species exhibited an excitation peak at a wavelength of 476 nm and an emission peak at a wavelength of 496 nm. In the presence of Ca^{2+}, its reconstituted aequorind isplayed an emission peak at 470 nm. It is also worth mentioning here that the photoactivity of the reconstituted aequorins of both *Aequorea victora* and *Aequorea.macrodatyla* were found to be similar. However, A. macrodactyla appeared brighter and more ``blue'' than *Aequorea victorea* because of the differences in the photoactivity of their GFPs [59].

It is reported that the margin of the umbrella of this species emitted light when the animal was touched [61]. The light was emitted in spots in or near the tentacular bulbs. Further, it was noticed that only the immediate region that was disturbed flashed and there was no spread of luminescence. Furthermore, the light was best from the base of the umbrella.

Aequorea victoria

Image credit: Sierra Blakely, Wikimedia

Aequorea Victoria (Fig. **5**) is a common species in the nearshore waters of the West coast of North America. It is a saucer-shaped hydromedusa with a maximum diameter of 18cm. It has 80 -100 narrow, unbranched radial canals which radiate from the central manubrium. The Bell of this species is thick and gelatinous. Its slender gonads which are bluish in males and rosy in females run along most of the length of the radial canals. There are about 150 tentacles which are in a single row around the margin of the bell. When not fluorescing, this medusa appears colorless and transparent.

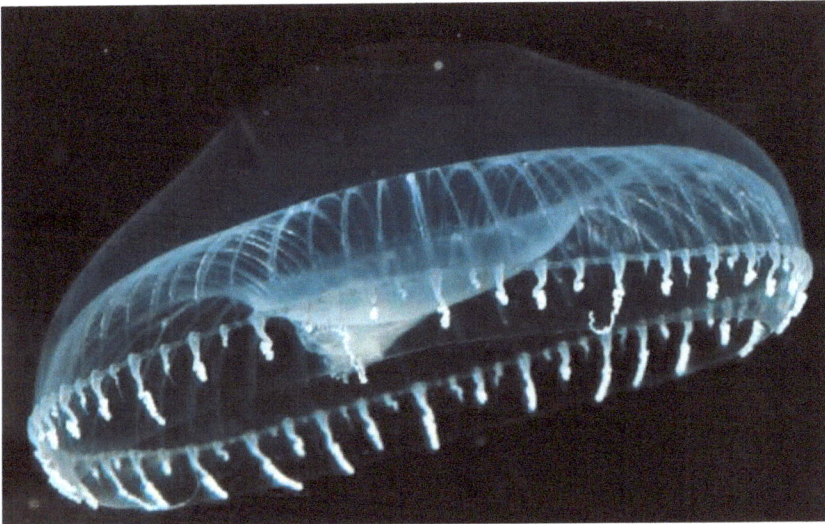

Fig. (5). *Aequorea victoria.*

Bioluminescence: In the bioluminescent specimens of this species, light is produced by its blue fluorescent, calcium-sensitive photoprotein Acquorin and associated green-fluorescent protein(GFP) present around the bell margin. Its flash or glow was green when distressed. In the light emitting process, the GFP transforms the blue light from Aequorin into green fluorescent light. Even when this species is shaken lightly, its photophores has been reported to give off their characteristic green glow. In the luminescent reaction, aequorin generates blue light with a broad emission peak with a wavelength of 460 nm, which directly excites GFP molecules to be re-emitted as green light with a higher wavelength of 508 nm [16].

Bougainvillia carolinensis

Image not available

The bioluminescent medusae of this species are common in Western Atlantic in lagoon and inshore areas at depths of about 3 to 10 m. Its marginal tentacles are arranged in four bundles. Oral tentacles are divided twice and 4 inter-radial gonads are seen. Umbrella is 4 mm wide and high and it is dome-shaped with very thick walls. Manubrium is long and narrow and marginal bulbs are small, bulbous, and each of them is with 7-9 slender, stiff tentacles. Ocelli are large.

Bythotiara depressa

Image not available

This bioluminescent medusa is a deep water, epipelagic species occurring at depths of about 400 m in the Eastern Pacific: USA and Canada.

Clytia hemisphaerica (= Clytia languida)

Image credit: Jccardenas13; Elizabeth Lee, Ph.D., Wikipedia

Clytia hemisphaerica (Fig. **6**) is a shallow water (about 20 m) eurythermic and euryhaline species found distributed in the Atlantic Ocean, Colombia to Argentina . The body of the medusa which has an initial diameter of only 2–3 mm is oval to linear and its ovaries extend distal half of the radial cana. There are 30 marginal tentacles and 1–2 statocysts between successive tentacles.

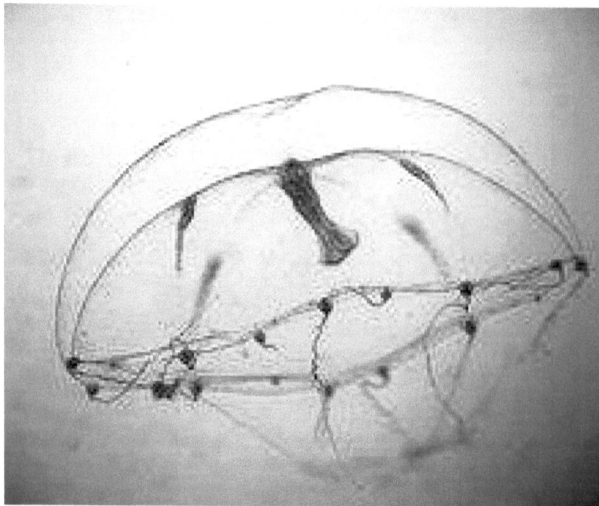

Fig. (6). *Clytia hemisphaerica.*

Bioluminescence: In this species, tiny light flashes were first recorded in their tentacle bulbs but no pigmented photophores were distinguishable. An additional dull glow from its gonads was also sometimes seen. Bioluminescence was also observed in the eggs, planula larva, and polyp form [16]. (Fourrage *et al*., 2014). It is reported on the manubrium + tentacular bulbs + tentacles fluorescence pattern in this species [64].

Clytia gregaria (= *Phialidium gregarium*)

Clytia gregaria (Fig. **7**) is one of the most abundant hydrozoans of the Northwest Pacific ocean particularly during spring and summer. The medusae showed an initial diameter of 2–3 mm and mature individuals may reach 2 cm in diameter. Its translucent bell is saucer-shaped, with a diameter greater than its height. There are four radial canals that are positioned close to the bell margin and are with a white or yellowish elongated gonad in each. On the bell margin of the mature individuals, there are up to 60 highly extensile tentacles of equal length. The mouth opens directly into the manubrium, which has four ruffled lips and is pale yellow or brownish. Rarely there is a lateral stripe of dark pigment on the bell margin and gonads.

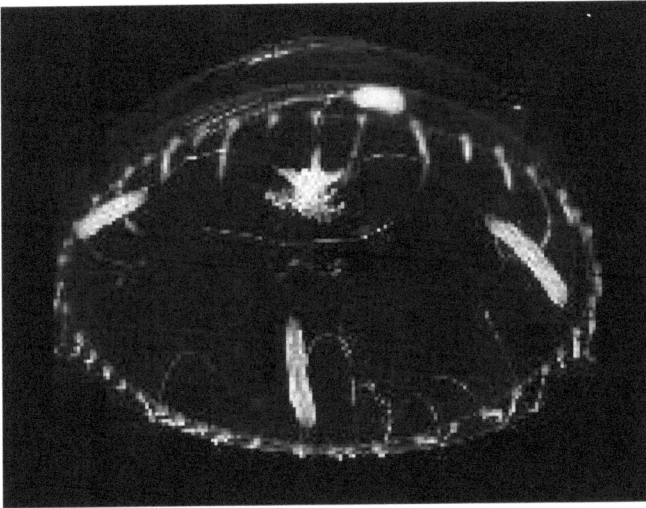

Fig. (7). *Clytia gregaria.*

Bioluminescence: In this species, the bioluminescence is confined to point sources around the bell margin. Emissions of all these medusae are blue [60]. The species of this genus possess a photoprotein, clytin (or phialidin), a relative of Aequorin family photoproteins [16]. The Green-fluorescent protein of this species (cgreGFP) has been reported to be useful intracellular fluorescent marker, as it was able to be expressed in mammalian cells [65].

Colobonema sericeum

Image credit: Mayor, Alfred Goldsborough, Wikimedia

This deep-sea hydromedusan species (Fig. **8**) is found distributed in the Antarctic, Atlantic, Indo-Pacific and Arctic Oceans. The umbrella of this silky medusa is up to 45 mm wide and 35 mm high. It is slightly conical and its manubrium is short and tubular. Gonads are linear and are located along the greater part of radial canals. There are 32 long tentacles and adradial tentacles develop before interradials.

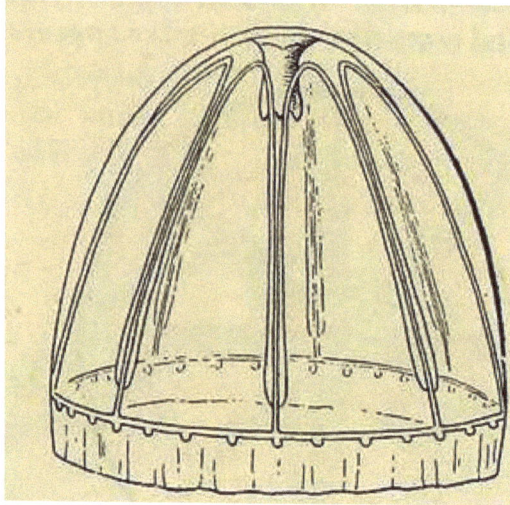

Fig. (8). *Colobonema sericeum.*

Bioluminescence: In this species, the bioluminescence is confined to point-sources around the bell margin. Additionally, the tentacles also produce light (bioluminescence), which is said to distract predators to make their escape. Emissions of all these medusae are blue [60]. It has been reported to produce a wave of light when touched. However, when vigorous attempts were made to capture the animal, it shed all parts of its tentacles as a luminous mass and its darkened bell swam off.

Cosmetira pilosella

Image credit: WoRMS

Medusae of this luminescent species (Fig. **9**) are largely found in the coastal to offshore waters of NE Atlantic Ocean. The umbrella of this luminescent species is somewhat flatter than the hemisphere and is fairly thick in the upper region thinning towards margin. Velum is broad. The stomach is short is quadrate with a broad base attached to subumbrella. The mouth is with four lips whose margins are crenulate. There are 4 narrow radial canals. Gonads present on radial canals are linear and 1/2 - 3/4 length of the radial canal. A maximum of 100 marginal tentacles which are e short with large round basal bulbs but without ocelli; There are 6-10 marginal cirri between adjacent tentacles and older cirri are found extending up onto the exumbrellar surface. The maximum diameter of an umbrella is 48 mm. While stomach and gonads are reddish violets, marginal tentacle bases are deep purple with a reddish centre.

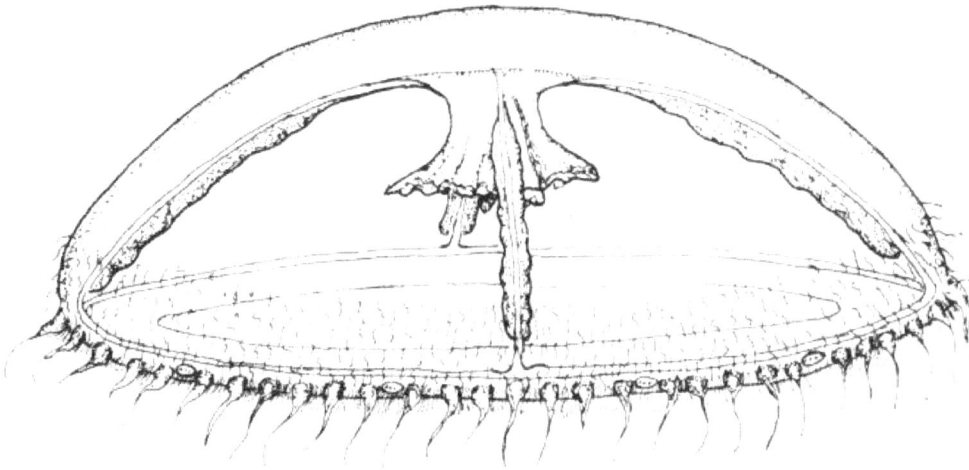

Fig. (9). *Cosmetira pilosella.*

Crossota alba

Image credit: Flickr

This bioluminescent species (Fig. 10) does not have a sessile stage as other hydromedusae and is found largely distributed in the Pacific ocean and west coast of India. The umbrella of this luminescent medua is up to 42 mm wide and 10 mm high. It is colorless; with 8 sharp longitudinal ridges which are separated by 8 broad, flat, furrows A maximum of 190 tentacles is present. Gonads are present on 8 radial canals and are nearer circular canal than manubrium.

Fig. (10). *Crossota alba.*

Cunina globosa

Image credit: WoRMS

Cunina globosa (Fig. **11**) is a deep water species inhabiting depths of about 400 m. Umbrella of this luminescent medusa is up to 18 mm wide and is almost globular. Manubrium is on broad peduncle. There are 10-14 tentacles that are arising slightly above the margin. Peripheral canals are narrow and marginal lappets are short and broad.

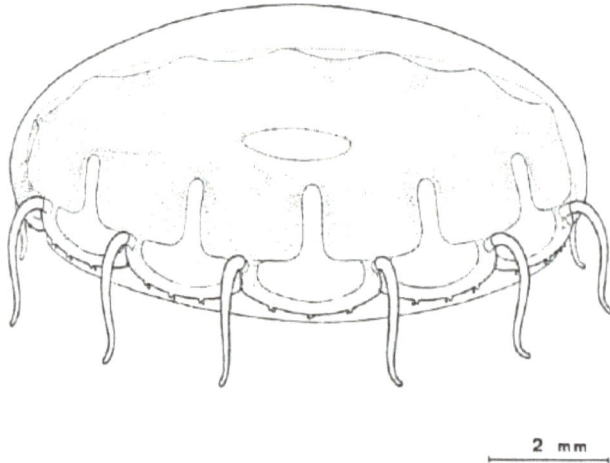

2 mm

Fig. (11). *Cunina globosa.*

Bioluminescence: In this species, the bioluminescence is confined to point-sources around the bell margin. Emissions of all these medusae are blue [60].

Eugymnanthea inquilina

Image not available

This hydroid species inhabits commonly in the bivalve hosts, *viz.* the mussel *Mytilus galloprovincialis* and the clam *Ruditapes decussatus* in the coastal waters of Italy and Spain of Mediterranean. The production of medusoids was high in mussels, whereas medusoids were rare and often abortive in clams. In its life cycle, the zygote develops into planula and later into polyp then into a planktonic medusa . The umbrella of this luminescent medusa is sphero-conical and is up to 0.6 mm high and wide, Manubrium and tentacles are absent. There are 4 simple radial canal that bear 4 large, sac-like gonads. There are also 8 adradial statocysts with 3 or 4 statoliths.

Bioluminescence: This medusa possesses umbrella margin fluorescence pattern. It is also reported bright green auto-fluorescence in the umbrella margin of a spent medusa of this species [64].

Eugymnanthea japonica

Image not available

The medusae of this species occur in Japanese waters. The mature medusa (1-da--old after release from the hydroid) of this species had a maximum diameter of 1.4 mm and it possessed eight statocysts, eight statoliths, and eight marginal warts.

Bioluminescence: In the subumbrella of a mature medusa of *Eugymnanthea japonica* from Japan, green fluorescence was observed. Bright green auto-fluorescence was detected in different parts of the body such as the umbrellar margin, umbrellar marginal warts, tentacular bulbs, tentacles, and manubrium of the laboratory-reared immature (1 –14 days old) medusae of *Eutima japonica* from Japan and China. Its fluorescence emission spectra with an emission maximum at 503 nm I showed the contribution of green fluorescent protein (GFP) [64].

Euphysora valdiviae

Image not available

It is a deep water species with luminescent medusae occurring at depths of more than 2100 m in Indonesia and West Sumatra only. Medusa is usually with 3 short or rudimentary marginal tentacles and one long, well-developed marginal tentacle. Umbrella is with 4 radial canals.

Eutima japonica

Image credit: Schuchert, Peter, WoRMS

This hydroid species (Fig. **12**) inhabits mainly on the soft body tissues of the juveniles of the Japanese scallop *Mizuhopecten yessoensis* in Japanese ((Northwest Pacific) waters. The medusae which are of 0.2 mm diameter of this hydroid are planktonic and luminescent [66].

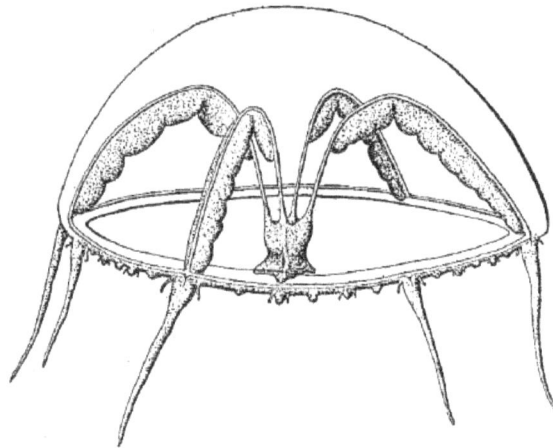

Fig. (12). *Eutima japonica.*

Bioluminescence: This luminescent medua has Manubrium + tentacular bulbs + tentacles + umbrellar margin fluorescence pattern [64].

Eutonina indicans

Image credit: Flickr

This hydroid species (Fig. **13**) occurs in the waters of Europe and Northern America. It has a transparent bell of only 25-35 mm diameter. Umbrella is slightly flatter than a hemisphere. Stomach is short and is situated on the elongate conical gastric peduncle. The mouth is with four folded lips. There are four radial canals which are extending along the gastric peduncle. Both radial and ring canals are narrowed. There is a large number of marginal tentacles. Stomach, gonads and marginal tentacle bases are white in color. Dark pigment is seen on tentacle bases and along dorsal surfaces of grooves in the roof of the stomach.

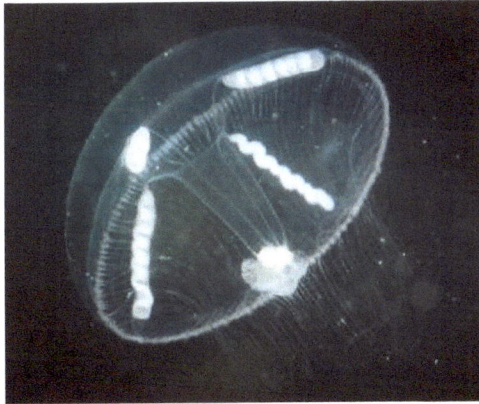

Fig. (13). *Eutonina indicans.*

Bioluminescence: Captive-reared luminescent medusae of this hydroid species produced flashes upon stimulation when they were directly injected with coelenterazine [67].

Halicreas minimum

Image credit: NOAA Office of Ocean Exploration and Research, Wikimedia

It is a deep sea, cosmopolitan hydrozoan (Fig. **14**) living at depths of about 1100 m. Umbrella of this bioluminescent medusae is 30-40 mm wide, thick and disk-like, with small apical projection. There are 8 clusters of gelatinous papillae above the margin. The mouth is a wide circular opening. There are 8 broad, band-like, radial canals and a broad circular canal. Gonads are flattened and are extending along almost the entire length of radial canals. Tentacles up to 640 are present and 3-4 statocysts are seen in each octant.

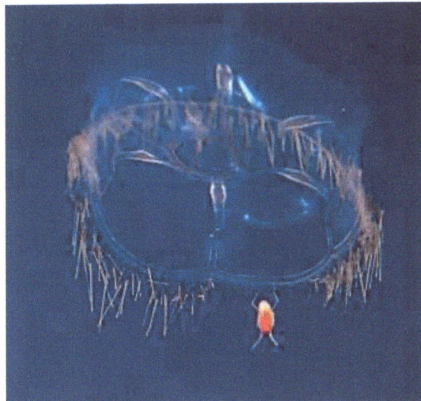

Fig. (14). *Halicreas minimum.*

Haliscera conica

Image credit: Schuchert, Peter, WoRMS

It is fairly a deep water species (Fig. **15**) living at a depth of about 400 m. Umbrella is up to 18 mm diameter with a very thick, bluntly conical projection. There are 64-72 marginal tentacles in adults and 8-9 tentacles and 2 statocysts in each octant. The base of tentacles is surrounded by a broad thickening of marginal cnidocyst tissue. Gonads are oval and are well separated from manubrium in the middle portion of 8 broad radial canals.

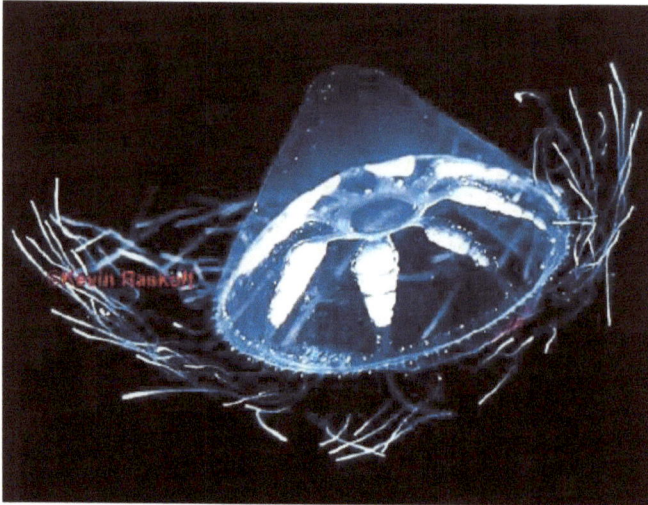

Fig. (15). *Haliscera conica.*

Bioluminescence: It has been reported to produce rapid propagated waves of light (from point sources) which spread over the aboral surface and round its umbrellar margin [68].

Halitrephes maasi *(= Halitrephes valdiviae)*

Image credit: NOAA Office of Ocean Exploration and Research, Wikimedia-

This deep-sea hydrozoan (Fig. **16**) is commonly called as firework jellyfish which occurs at depths of more than 700m in warm and temperate regions. Umbrella of this medusa is 100 mm wide and its mesoglea is flaccid. There are 16-30 broad, ribbon-like radial canals, in which some of them are bifurcated. A total of 100-300 marginal tentacles are present.

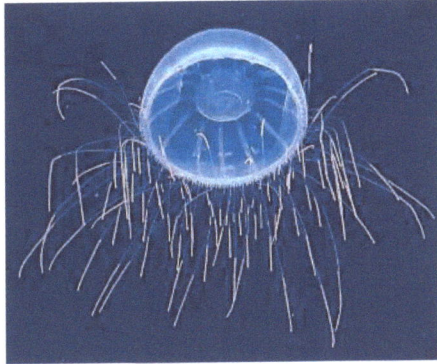

Fig. (16). *Halitrephes maasi.*

Bioluminescence: Its radial canals have been reported to form a starburst pattern that reflects the lights of ROV [remotely operated underwater vehicle] Hercules with bright splashes of yellow and pink [69].

Halopsis ocellata

Image credit: WoRMS

Halopsis ocellata (Fig. **17**) is a common shallow-water species living at depths of only 20 m. However, it has been reported from both off-shore as well as in the coastal areas of eastern and western parts of the North Atlantic Ocean. The luminescent medusa stage of this species has an umbrella that is approximately 1/4 of the diameter. The diameter of a mature medusa is 6-7 cm. Its e stomach is 1/5 of the diameter, circular or star-shaped. There are 11-17 radial canals. Gonads form folded bands along two-thirds of each canal and are greyish, pinkish, or white. There are 450 radial canals.

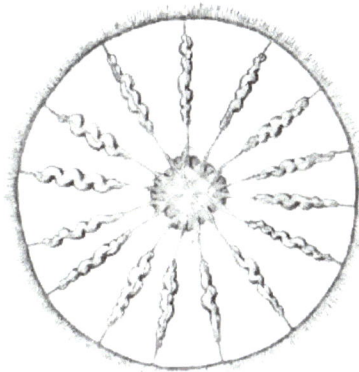

Fig. (17). *Halopsis ocellata.*

Bioluminescence: It emits both blue and green light and the latter is due to its GFP content [57].

Leuckartiara octona

Image credit: Schuchert, Peter, WoRMS

This luminescent medusa (Fig. **18**) occur at a depth range 0 - 210 m in Atlantic Ocean, Mediterranean, Sub-Antarctic and temperate waters of the northern hemisphere. Umbrella which is up to 20 mm or more high.is bell-shaped, higher than wide, with well-developed conical or spherical solid apical projection. Its four radial canals are broad and ribbon-like and are with smooth or slightly jagged edges. Crenulated mouth lips are seen. Manubrium and marginal tentacles are pink to crimson or yellowish brown. In coloration.

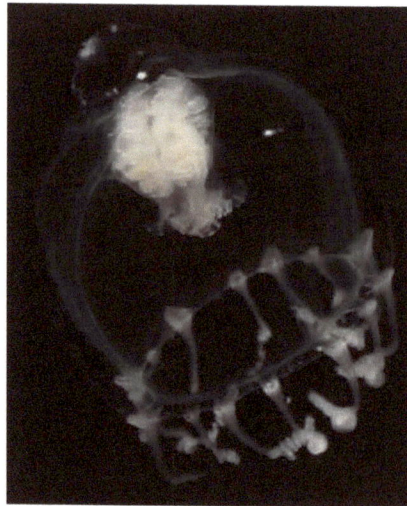

Fig. (18). *Leuckartiara octona.*

Liriope tetraphylla

Image credit: Daiju Azuma, Wkimedia

This luminescent medusa (Fig. **19**) lives in the epipelagic zone of the warm and tropical seas such as North Sea, North, East, and West South Atlantic Ocean, Mediterranean, and in Indo-Pacific Oceans. Medusae are transparent and colourless and the most obvious characters of these organisms are the dome-shaped bell and very long gelatinous peduncle with a very small stomach at the extreme end. Inside the bell, there are four broad leaflike gonads that are slightly greenish or pinkish reflective sheen. It has four tentacles, one opposite each gonad. Bell diameter is up to 30mm.

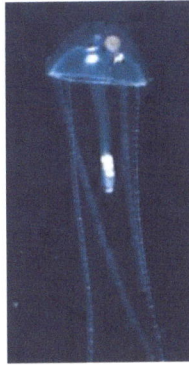

Fig. (19). *Liriope tetraphylla.*

Bioluminescence: It is known for its green fluorescence and a tentacular tip fluorescence pattern is seen in this medusae [64].

Mitrocoma cellularia *(= Halistaura cellularia)*

Image credit: David Young (Reproduced with permission)

Mitrocoma cellularia (Fig. **20**) is a shallow water hydromedusa (Fig. **20**) occurring in a depth of about 5 m. There are 100-350 unbranched, filiform tentacles originating at the margin of the bell and are evenly distributed. Long tentacles alternate with short tentacles. Manubrium which has four long and extended lips is attached directly to the subumbrella and is not suspended below the bell margin by any peduncle. Its 4 radial canals do not branch without any major diverticula . Gonads are linear and are associated with the radial canals. These medusae are colorless and their gonads and exumbrella are pale blues. Bell's diameter is 4 cm.

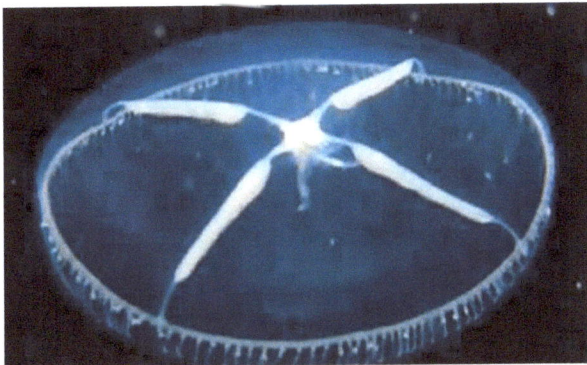

Fig. (20). *Mitrocoma cellularia.*

Bioluminescence: A narrow bioluminescent band circles the edge of the body and the hundreds of white tentacles originating from the bell can glow with a blue-green coloured light [70]. The photocytes of the outer bell margin of this species have been reported to possess the photoprotein mitrocomin (an Aequorin family photoproteins) with bioluminescence properties. Although this mitrocomin isoform revealed a high degree of identity of amino acid sequences, they are said to vary in specific bioluminescence activities. That is, all isotypes displayed the identical bioluminescence spectra (473-474nm) [16, 71].

It is reported that the luminescence of this species was observed as a continuous ring of light around the margin of the bell of this species. Its emissions were blue, with emission maxima between 460 and 494 nm [60]. Further, this species has been reported to possess green fluorescent protein (GFP) [57].

Mitrocomella polydiademata

Image credit: Schuchert, Peter, WoRMS

The umbrella of this luminescent medusa (Fig. **21**) is hemispherical and jelly is fairly thick. Velum is prominent and is about 1/4 bell radius. The stomach is short, and four-sided. The mouth is with four short lips with slightly folded margins. Its 4 radial canals and ring canal are narrow. The diameter of umbrella in mature medusae is 9-30 mm. While its stomach and marginal tentacle bases are purplish or rosy, its gonads are yellowish-brown, purplish or rosy.

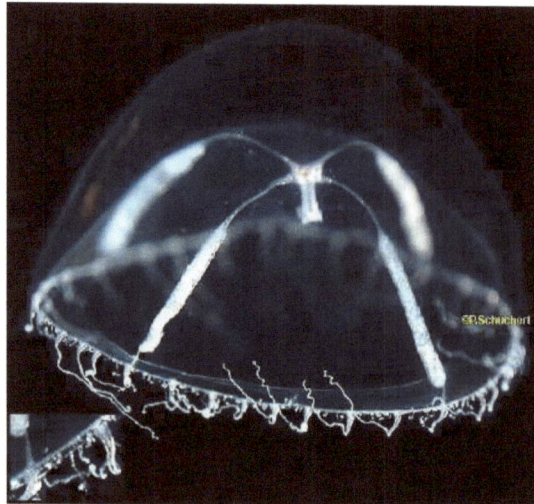

Fig. (21). *Mitrocomella polydiademata.*

Obelia dichotoma

Image credit: Dave Cowles (Reproduced with permission)

The luminescent medusae of this species (Fig. **22**) is commonly distributed in Alaska to San Diego, and worldwide in temperate zones. It is a small hydromedusa with a disk-shaped bell with a diameter of up to 4 mm. It has no ocelli. Manubrium is short. Its 16 unbranched tentacles are of the same size, and are distributed evenly along the margin of the bell. Tentacles do not contain rings of nematocysts. Its 4 radial canals do not branch and are without major diverticula. Gonads are almost spherical and are attached about halfway along the radial canals. There are 8 statocysts.

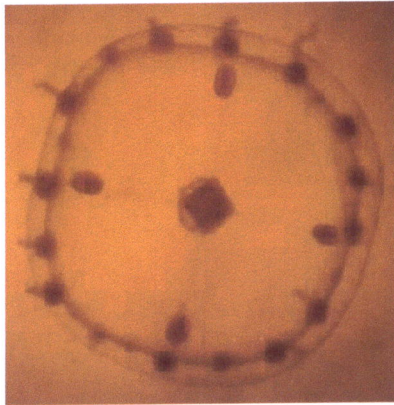

Fig. (22). *Obelia dichotoma.*

Bioluminescence: Its green fluorescence is from its tentacular bulbs.

Obelia sp. aff. *dichotoma*

Image not available

Bioluminescence: This species emits green fluorescence and it is from its manubrium and tentacular bulbs [64].

Obelia lucifera

Image credit: Valerie Jane Morse

Obelia lucifera is the luminous medusa (Fig. **23**) from *Obelia geniculata* [72]. The medusa stages of *Obelia* spp. are very common in the coastal and offshore plankton throughout the world. Umbrella of *Obelia* medusae is flat. The stomach is short and is with quadrangular base. The mouth is with four short simple lips.

There are four radial canals that are straight and narrow. Gonads are spherical to ovoid and sac-like and are hanging from middle to end-regions of radial canals. Numerous short solid marginal tentacles are seen.

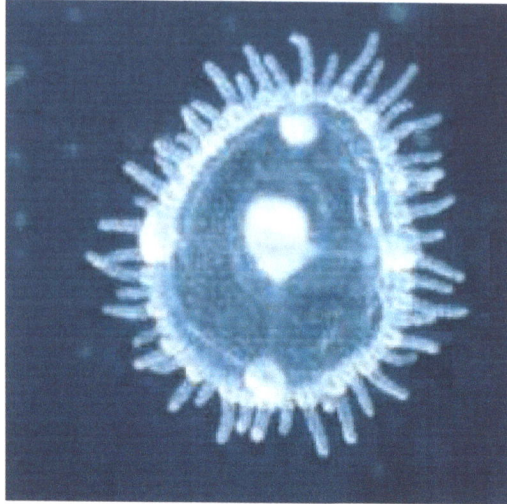

Fig. (23). *Obelia lucifera.*

Bioluminescence: The tiny photocytes located around the rim of this medusa's umbrella and in the transparent tentacles produce green fluorescence. All the medusae of *Obelia* possess the Aequorin family photoproteins *viz.* obelins which are calcium-sensitive and are involved in the emission of blue light [16]. Obelin contains coelenterazine linked to apo obelin by a covalent bond. Within the obelin complex coelenterazine is in its peroxidised form and acts as the most reactive part of the photoprotein. Its green-fluorescent protein (GFP) functions to modify the emission wavelength of its bioluminescence system and converts the blue light into green [17]. The maximum wavelength (λ max) of obelins was identical with that of the photoproteins from *Obelia geniculata* and *Obelia longissimai.e.* 475 nm [72].

Octophialucium funerarium

Image credit: WoRMS

This luminescent medusae (Fig. **24**) are found living at depths of 15-200 m In the Western Indian Ocean, Northeast Atlantic, Arctic Sea, and the Mediterranean. The umbrella of this medusa has eight radial canals and adaxial excretory papillae There are no ocelli. Its eight gonads are completely surrounding the radial canals and are separated from the stomach. There are also eight mouth lips.

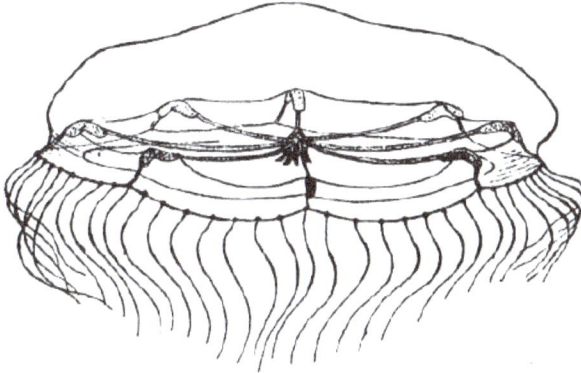

Fig. (24). *Octophialucium funerarium.*

Olindias formosus (=Olindias formosa)

Image credit: Fred Hsu, Wikimedia

These flower hat jellies (Fig. **25**) occur in the northwestern Pacific off central and southern Japan, and South Korea. The adult flower hat jelly has been reported to live only for a few months from December to July, with peaks in April and May. It has lustrous tentacles that coil and adhere to its rim when they are not in use. The bell of this luminescent medusa is translucent and pinstriped with opaque bands. This jelly can grow to a maximum diameter of 15 cm.

Fig. (25). *Olindias formosus.*

Bioluminescence: These medusae possess green fluorescent bands near the tips of their upturned tentacles, with a blue-absorbing chromoprotein at the tentacle tip. Illumination of these medusae was found alternated between blue, yellow, and white LEDs, (light-emitting diodes). The blue LED produced a maximum emission at 479 nm and the yellow LED at 566 nm [17].

Pandea conica

Image not available

It is a deep water species living at depths of 450 m in the North-East Atlantic, the North Sea, the English Channel and the Mediterranean Sea. The umbrella of this luminescent medusae is up to 10 mm wide and 30 mm high, with rounded or conical apex. A yellow-orange border line runs around the umbrella. There are 16-24 longitudinal exumbrellar cnidocyst ribs which are originating from each tentacular bulb. Manubrium is with short mouth tube and very folded lips. Its radial canals are fairly narrow and smooth. Marginal tentacles are with conical, laterally compressed bulbs. Its bright red gonads are located on entire interradial walls of manubrium, forming a coarse-meshed network of ridges with pits in between.

Pegantha clara

Image credit: Kevin Raskoff, MBARI. WoRMS

Pegantha clara (Fig. **26**) is found distributed in Antarctic Atlantic, Eastern Pacific and the Atlantic Ocean at depths of 200-300m where maximum density has been observed. Umbrella of this luminescent medusae is up to 50 mm wide and 20 mm high. There are 20-40 quadrate marginal lappets, each with 3-5 statocysts. Gonads are somewhat crenulated.

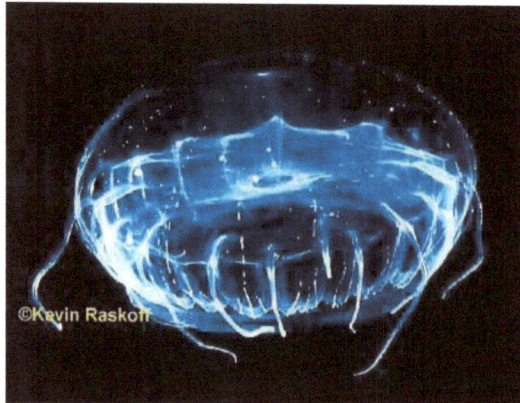

Fig. (26). *Pegantha clara.*

Pegantha laevis

Image credit: Russell R. Hopcroft (University of Alaska Fairbanks, USA) (Reproduced with permission)

This luminescent medusae (Fig. **27**) are found widely distributed in tropical, temperate and even Antarctic) regions of the three large oceans at depths of about 300 m. Umbrella is flat, lenticular and is up to 40 mm wide. Its marginal lappets (16-26) are square, each with 5-7 statocysts. Ex-umbrella is smooth. Its gonads are sac-shaped when fully developed, with oval processes.

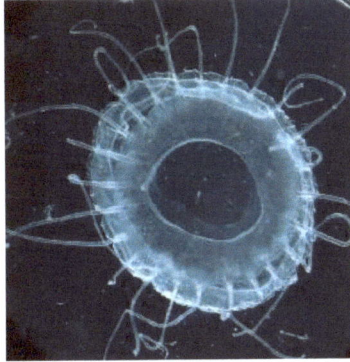

Fig. (27). *Pegantha laevis.*

Proboscidactyla ornata

Image credit: Flickr

Proboscidactyla ornata (Fig. **28**) is a cosmopolitan and euryhaline species commonly occurring in the tropical Atlantic Ocean and Caribbean waters at a depth range of 0 - 85 m .The umbrella of this luminescent medusa is hemispherical with a maximum diameter of 4 mm and is with 16 hollow, marginal tentacles Apical region of the umbrella has thickened mesoglea. The base of its manubrium is with four gastric pouches, Mouth has four lips. There are four primary radial canals that end in terminal branches. Tentacular bulbs are orange to brown.

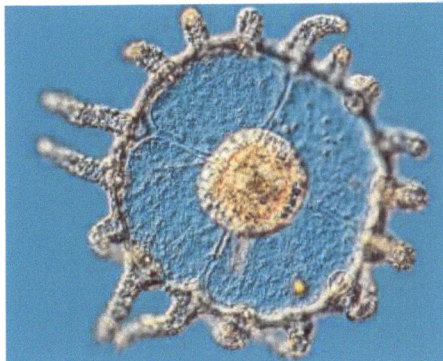

Fig. (28). *Proboscidactyla ornata.*

Bioluminescence: its subumbrella and manubrium have been reported to emit green light [64].

Rathkea octopunctata

Image credit: Flickr Image credit: De Blauwe, Hans, WoRMS

Rathkea octopunctata (Figs. **29** and **30**) is mostly a boreal species but it may also be found in subtropical and tropical Indo-Pacific, Atlantic, Arctic and the Mediterranean. Bell height of this luminescent medusae measures 4 mm. It has a short manubrium on a gelatinous peduncle. There are 4 distinct lips on the mouth and each is with a pair of separated nematocysts knobs. Four radial canals and 8 tentacle bulbs are seen. A maximum of 5 tentacles are present on each bulb at the base of each radial canal and 1-3 tentacles on the alternate bulbs. Tentacle bulbs and stomach are yellowish to reddish or brown in coloration

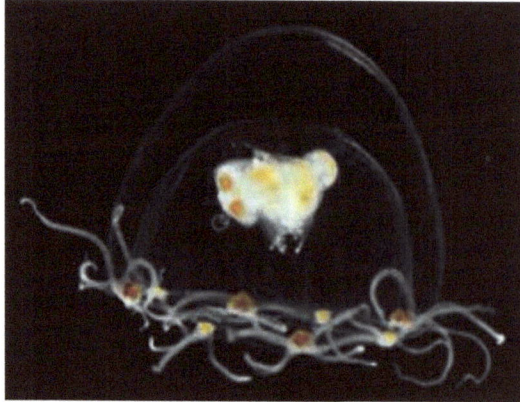

Fig. (29). *Rathkea octopunctata.*

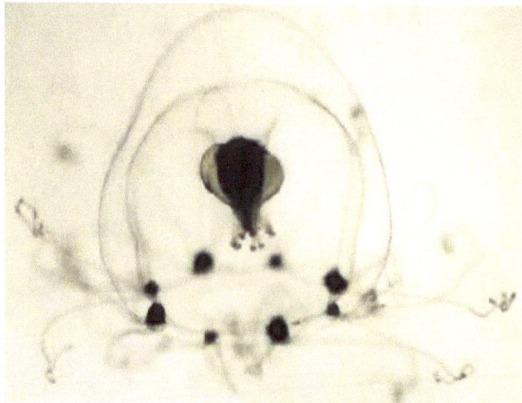

Fig. (30). *Rathkea octopunctata.*

Bioluminescence: Its manubrium and tentacular bulbs have been reported to emit green light [64].

Solmaris leucostyla

Image credit: Josep-Maria Gili (Reproduced with permission)

Solmaris leucostyla (Fig. **31**) is a Mediterranean species and its luminescent medusae possess dome-shaped bells which bear numerous tentacles arranged on the undulating margin of the bell. The characteristic gastric pouches are absent in these medusae. Gonads are situated inside the wall of the stomach.

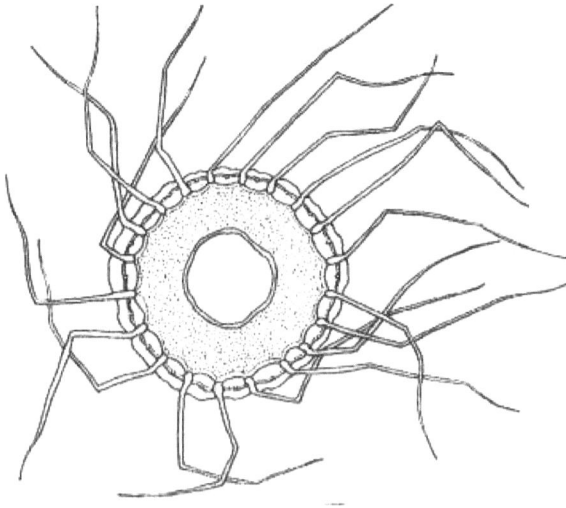

Fig. (31). *Solmaris leucostyla.*

Bioluminescence: Very small (2-3 mm) medusae of this species produced bioluminescent responses similar to those of other species of *Solmissus*. The species of *Solmissus* are known to emit a continuous ring of blue light around the margin of their bells with emission maxima between 460 and 494 nm However, these medusae have been reported to significantly contributor to the luminescence in near-surface waters because of its numbers [68].

Solmissus albescens

Image credit: Christian Coudre (Reproduced with permission)

Solmissus albescens (Fig. **32**) is a deep water species occurring in Mediterranean: Morocco and Monaco at depths of about 500 m. Its luminescent medusa has a transparent and discoid umbrella which is often found stiffened by a fine chitinous envelope. In these medusae, only their velum contract for their movements. Maximum diameter of the bell is 3 cm. There are 16 rigid and non-retractable whitish tentacles rooted on the edge of the bell. These tentacles are found radiating indifferently below or above the plane formed by the discoid bell. These medusae have been reported to make vertical migrations by going to depth during the day and s approaching the surface at night.

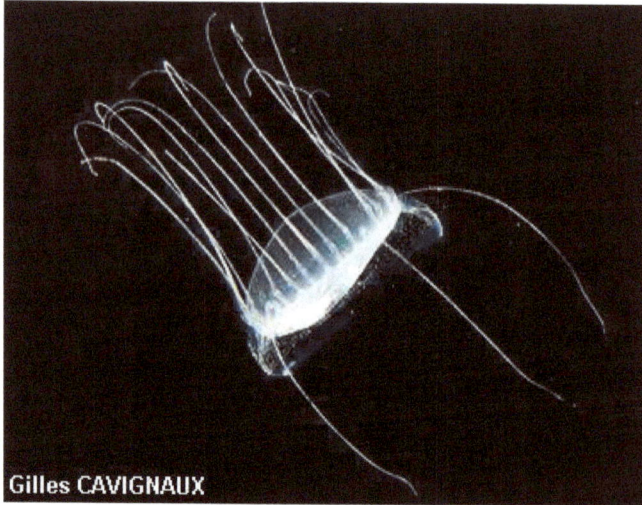

Fig. (32). *Solmissus albescens.*

Bioluminescence: The photosensitive cells (ocelli) of these medusae have been reported to detect variations in light and allow them to correlate their migrations with the day-night cycle [73].

Solmissus incisa

Image credit: Jian Rzeszewicz CC

Solmissus incisa (Fig. **33**) is a neritic and deep water species occurring at depths of 0-1000 m in the Atlantic, Pacific Oceans, Mediterranean and Antarctic sea. Its luminescent medusa is with a flattened, disc-like exumbrella which is concave at the apex. Mesoglea is fairly thick but transparent and soft. Velum is well-developed. The manubrium is large and circular. Twenty-two to twenty-four perradial manubrial pouches is seen. There are also 22- 24 marginal tentacles that are whitish at their tips presumably due to very high concentrations of nematocysts.

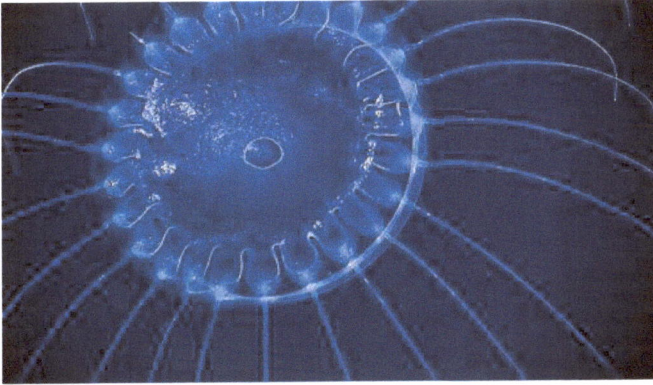

Fig. (33). *Solmissus incisa.*

Solmissus marshalli

Image credit: Flickr

Solmissus marshalli (Fig. **34**) is fairly a deep water species occurring at a depth of 250 m in Western Indian, Atlantic and Pacific Oceans. The medusae of this species are commonly called as dinner plate medusae which are completely transparent and colourless. The exumbrellar surface of these medusae is completely smooth. There are 8-20 rectangular stomach pouches which are longer than wide. Many fine long tentacles are seen on the radii.

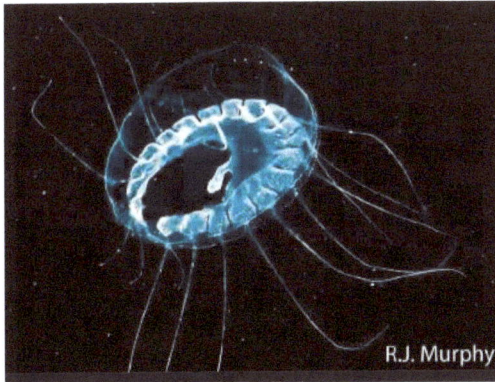

Fig. (34). *Solmissus marshalli.*

Bioluminescernce: It is reported that the luminescence of this species was observed as a continuous ring of light around the margin of the bell of this species. It emitted blue light, with emission maxima between 460 and 494 nm [60]. The medusae of the species of Solmissus emit bright light from their lateral umbrellar margin, the velum and down the tentacles [68].

Solmundella bitentaculata

Image credit: Aino Hosia/Universitetsmuseet i Bergen, Universitetet i Bergen, Wikimedia

The luminescent medusae of this species (Fig. **35**) occur at depths of 0 - 1000 m in Arctic, Indo-Pacific, Atlantic, Mediterranean and the Antarctic. It is also a euryhaline species. Umbrella is hemispherical and its margin is lobed. Bell is up to 12 mm in diameter. Apical mesoglea is very thick. Tentacles arise from inside the animal and not from the margin. There are no secondary tentacles, tentacular bulbs, radial canals or benthic polyps. Gonads are present on the manubrium and the velum is broad. There are 8 manubrial pouches and 8-32 statocysts.

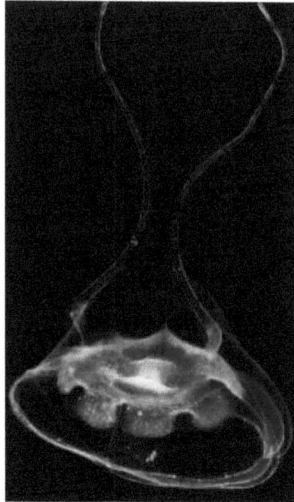

Fig. (35). *Solmundella bitentaculata.*

Spirocodon saltatrix (= *Spirocodon saltator*)

Image credit: Flickr

The body of this luminescent medusa (Fig. **36**) which is otherwise called hairy jelly is pyramidal, with a thick conical apical jelly. Apical mesoglea is very thick. There are two rigid tentacles that are inserted into the body near the apex, well above the bell margin. Eight manubrial pouches and 8-32 statocysts are present. There are no peripheral canal system, secondary tentacles or otoporpae. The maximum diameter of the medusa is 20mm and height is 6 cm.

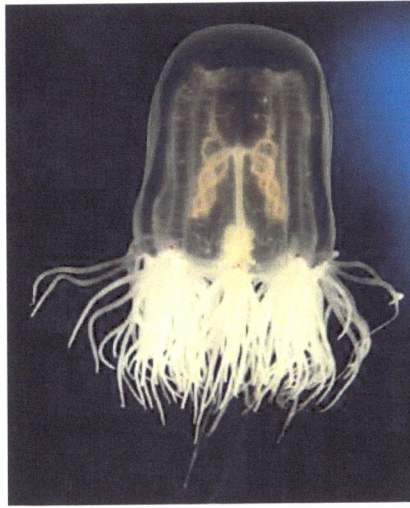

Fig. (36). *Spirocodon saltatrix.*

Bioluminescence: This medusa has been reported to emit light from its manubrium, tentacles, radial canals and umbrellar margin . Further, it also has subumbrellar fluorescence pattern [64]. In an experiment, when the light illuminating this medusa is turned off or is dimmed, this animal was found to respond to the decrease of light intensity by the pulsation of its bell. For this shadow reflex, its red-coloured ocelli present in the tentacles are believed to be responsible [74].

Tima bairdi

Image credit: Russell, https://ns-zooplankton.linnaeus.naturalis.nl/linnaeus_ng/ app/views/species/taxon.php?id=131812&epi=210

Tima bairdi (Fig. **37**) is a shallow water species occurring at a depth range of 0 - 20 m in Germany, Denmark and Scotland of the Northeast Atlantic Ocean. The umbrella of this luminescent medusa is hemispherical to slightly higher and is with a diameter of 6.5 cm. Jelly is very thick and velum is narrow, less than 1/4 of bell radius. The stomach is small, four-sided and is attached to a conical gastric peduncle by a cross-shaped base. The mouth is located completely outside the umbrella, and is having four large pointed lips with much-folded margins. Radial canals are four in number and are extending from bell edge to end of the peduncle. Both radial canals and ring canal are narrow. The color of marginal tentacles is pale pink or faint brownish at base and gonads are milky-white.

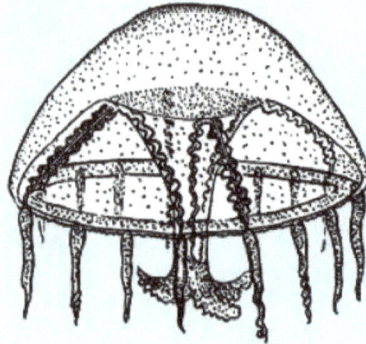

Russell, 1963d

Fig. (37). *Tima bairdi.*

Tima nigroannulata

Image credit: Dale Calder (Reproduced with permission)

The umbrella of these medusae (Figs. **38 - 40**) are usually higher than hemisphere and are with regular curvature. Its diameter at the margin is 23–46 mm and height, 12–38 mm. Mesoglea is thick at apex and is gradually thinning towards periphery. The gastric peduncle is well developed and conical. Tentacles are varying in number from 30 to 53. Statocysts are oval to almost spherical and are 71–141 in number. Manubrium located at tip of gastric peduncle is short and wide with bulbous pouches. The mouth is cruciform, with four perradial channels leading into 4 lips. There are 4 radial canals that are slender. Velum is thin, with a wide velar opening. Gonads are ribbon-like and are occurring continuously along radial canals from the tip of peduncle to the circular canal [75].

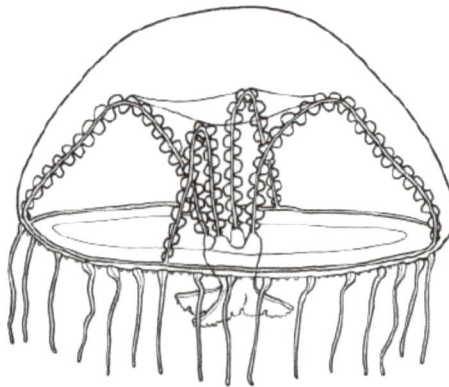

Fig. (38). *Tima nigroannulata (*Adult medusa).

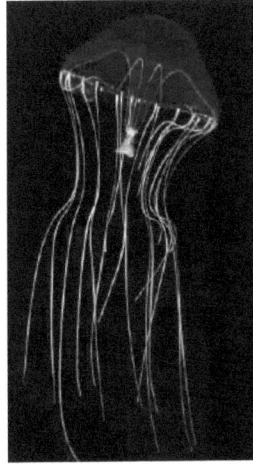

Fig. (39). Advanced juvenile medusa Fig. (40). Fluorescent Juvenile medusa.

Bioluminescence: When exposed to black light, the bases of tentacles of these medusae have been reported to glow light blue [75]. However, there is no evidence yet to indicate that this species is bioluminescent.

Tima saghalinensis

Image credit: WoRMS

Tima saghalinensis (Fig. **41**) is a dominant medusa in the central Sea of Okhotsk, Russia and Japan in the Northwest Pacific. The umbrella of the medusae of this species is much flattened. There are 250-300 tentacles.

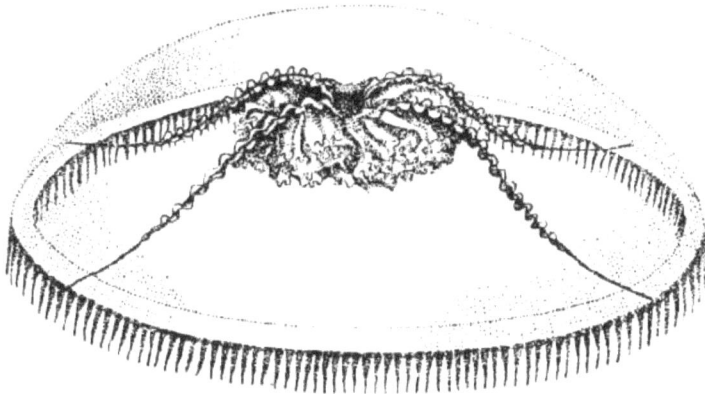

Fig. (41). *Tima saghalinensis*.

Bioluminescence: This species emits a continuous ring of light around the margin of the bell. It emitted blue light with emission maxima between 460 and 494 nm [60].

Luminescent Marine Siphonophores

Abylopsis eschscholtzii

Image credit: Riek, Denis, WoRMS

Abylopsis eschscholtzii (Fig. **42**) is a coastal, cosmopolitan species commonly distributed in warm/temperate seas. Dorsal and ventral facets of the anterior nectophore of this species are relatively more pentagonal and are3 of nearly equal size. Strongly serrated ridges are present. Lateral radial canals on nectosac are without ascending loop. The posterior nectophore is less than twice as long as wide with large apical apophysis. The dorsal facet of the cuboidal bract of this species forms a regular pentagon and apico-lateral facets are rectangular.

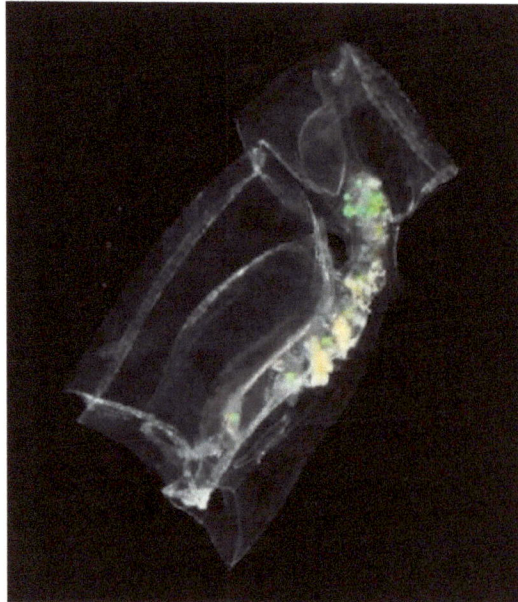

Fig. (42). *Abylopsis eschscholtzii.*

Bioluminescence: A two-domain green FP (abeGFP) and a four-domain orange-fluorescent FP (Ember) have been isolated from this species. The heterologously expressed siphonophore protein abeGFP showed a single excitation peak at 502 nm and a single emission maximum at 517 nm [58].

Abylopsis tetragona

Image credit: Riek, Denis, WoRMS

Abylopsis tetragona (Fig. **43**) is a shallow water species living at depths of 0-200 m in the Pacific Ocean, Atlantic Ocean, Antarctic and the Mediterranean. Animals are delightfully boxy and angular. The dorsal and ventral facets of the anterior nectophore are slightly pentagonal. Ridges are less strongly serrated. Lateral radial canals on nectosac are with ascending loop. The posterior nectophore is three times as long as wide. Two conspicuous basal teeth are seen. The dorsal facet of the bract is elongated and is less pentagonal. Apico-lateral facets are trapezoidal.

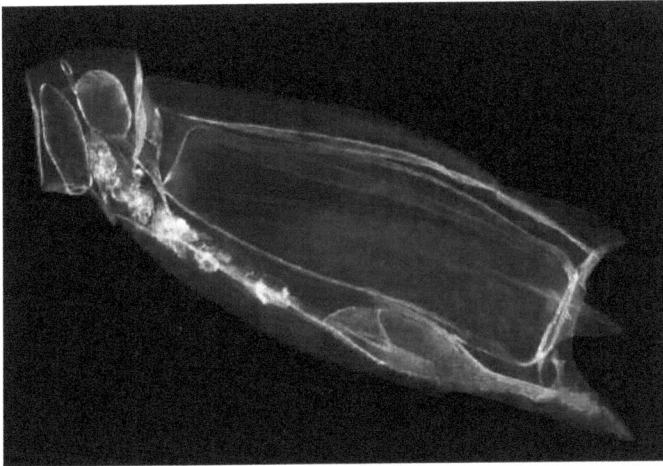

Fig. (43). *Abylopsis tetragona.*

Bioluminescence: The spectrum of this species was bimodal, with two distinct narrow peaks. These differences are associated with their specific body regions [57].

Agalma okeni

Image credit: Flickr

This epipelagic, tropical and subtropical species (Fig. **44**) occurs at a depth range of 0 - 200 m in the Pacific Ocean and Atlantic Ocean. In this luminescent species, apico-lateral ridges on nectophores are without distinct notch.. Lateral ridges are absent. Nectosac is Y-shaped in larger nectophores. Lateral radial canals are distinctly looped. Bract is with 4 characteristic distal facets. The Bracteal canal often ends before reaching the distal extremity.

Fig. (44). *Agalma okeni.*

Amphicaryon acaula

Image credit: Valerie Allain (SPC) (Reproduced with permission)

This species (Fig. **45**) occurs in the Antarctic, Atlantic, Pacific Oceans and the Mediterranean.These little, bioluminescent spheres of clear jelly are with yellow in the middle. Its sphere is comprised of two parts, one much larger and partially enclosing the smaller. The yellow bit inside is the stem. The diameter of this species may be up to 10mm. Its reduced nectosac has no ostial opening and the radial canals are simple, but are distinct.

Fig. (45). *Amphicaryon acaula.*

Amphicaryon ernesti

Image credit: Michael Boyle, Google Arts and Culture

This luminescent, tropical to polar species (Fig. **46**) occurs at a depth range of 0-200 m in the Pacific Ocean, Atlantic Ocean and Antarctic. Nectosac is found greatly reduced and is of a characteristic shape. Its ventral canal forms a network on the ventral wall, while the dorsal canal is simple. There are no lateral canals.

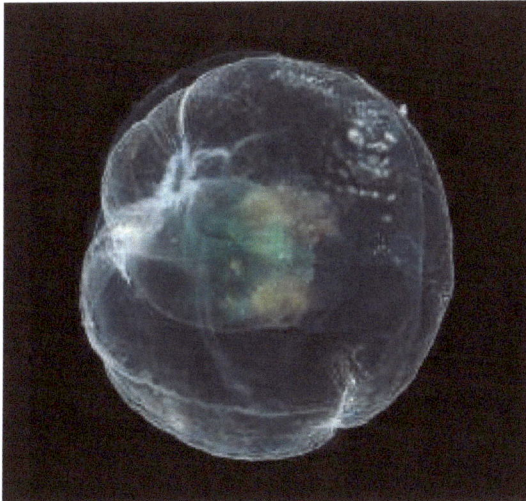

Fig. (46). *Amphicaryon ernesti.*

Apolemia sp.

Image credit: Catriona Munro, Stefan Siebert, Felipe Zapata, Mark Howison, Alejandro Damian-Serrano, Samuel H. Church, Freya E.Goetz, Philip R. Pugh, Steven H.D.Haddock, Casey W.Dunn, Wikimedia

The species (Fig. **47**) of this genus are considered to be the longest animal species on the planet, There are two distinct general patterns of siphosomal organization in different Apolemia species. In the first pattern *viz.* dispersed organization, zooids independently attach directly to the siphosomal stem. In the second pattern *viz.* pedunculate organization, only the gastrozooid is attached directly to the stem, and the other zooids of the cormidium branch from its peduncle. The size of this siphonophore's outer ring was found to be 154 feet long based on its diameter.

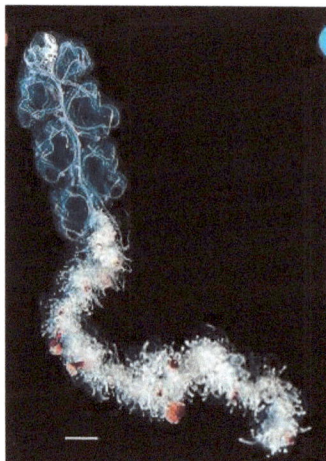

Fig. (47). *Apolemia* sp.

Bioluminescence: In the species of *Apolemia*, the bioluminescence of its nectosome orglnated from the stem rather than the nectophores, and this light propagated as bands of the fight at 50 cm s^{-1}. Initially, a flash ws found to propagate from the distal tip to the middle of the nectosome stem and it is followed by a series of flashes from the middle to the distal and proximal ends of the stem simultaneously, and it is ended with single flashes that alternately propagated from distal to middle and middle to distal portions of the stem. The propagated flashes in the nectosome stem lasted for approximately 2 s. Bioluminescence from the siphosome on the other hand lasted for more than 3.5 s [60].

Bargmannia elongata

Image not available

It is a subtropical species found in countries of Antarctic, Atlantic and Pacific Oceans *viz.* Brazil, Canada and USA. It has characteristically shaped nectophores which are elongated with large triangular thrust block. Its long, narrow nectosac has straight radial canals. Bract is broad and rounded with many patches of cells on its dorsal surface.

Bioluminescence: *Bargmannia elongata* did not display two distinct peaks, but a broad spectrum whose emission max varied by up to 56 nm (443 to 499 nm), which is 30% more than the entire range of the phylum Ctenophora. This variability is extreme enough to enable this species to represent both the bluest and greenest siphonophore spectra. Further, in this species, the peak of a broad unimodal spectrum moved from one end of the range to the other [57].

Bassia bassensis. (= *Abyla pentagona*)

Image credit: Riek, Denis, WoRMS

This epipelagic, temperate species (Fig. **48**) is found distributed in the Pacific, Indian and Atlantic Oceans, and the Mediterranean Sea. It is a luminescent monotypic siphonophore with a height of 38 cm and width of 27 cm. The animal of this species Includes colonies with swimming bells (nectophore) but the float (pneumatophore) is absent. Its anterior nectophore of this sepcies is polyhedral. Nectosac extends above the main body of the somatocyst. Its hydroecium does not extend up between the nectosac and somatocyst which is large, globular and without an apical diverticulum. Posterior nectophore is rectangular, with 4 main ridges which end in short basal teeth. Bract has a quadrilateral dorsal facet. Phyllocyst is a long tube which is swollen apically and is without lateral processes. Gonophore has 4 longitudinal ridges which end basally in minute teeth. A bluish tinge is seen to the ridges of both the nectophores and the eudoxids.

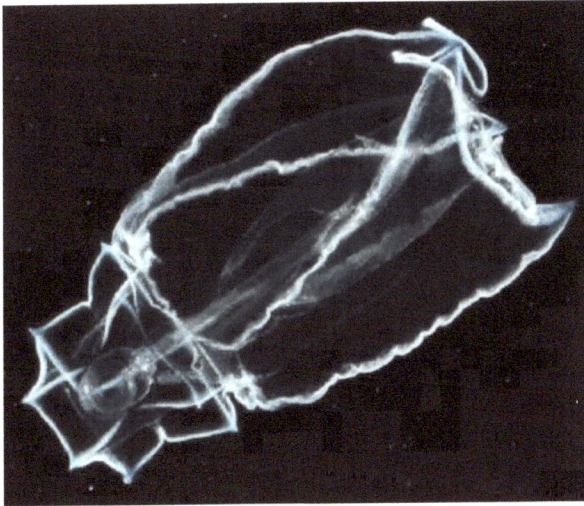

Fig. (48). *Bassia bassensis.*

Chelophyes contorta

Image not available

This epipelagic, subtropical species occurs at a depth range of 0 - 200 m in Indo-Pacific, Atlantic Ocean and the Mediterranean. In this luminescent species, the ventral facet of the anterior nectophore is strongly twisted to the right. Somatocyst is also twisted to the right. Posterior nectophore does not possess a distinct notched tooth on the ventral ridges.

Chuniphyes multidentata

Image not available

It is a deep- sea species found distributed in the Atlantic, Pacific Oceans and the Antarctic at a depth range of 0- 2500m. In this luminescent species, its characteristic feature *viz.* the somatocyst of the anterior nectophore is laterally expanded. Further, in its anterior nectophore, a discrete junction is present between the swollen region of the somatocyst and its thin anterior branch; a bifurcation is seen on the lateral ridge at 4/5 nectophore length; and there are a number of pointed cusps in both nectophores.

Craseoa lathetica

Image not available

This luminescent, fairly deep-sea species is found distributed in Western Atlantic *viz.* USA and Canada at depths of about 500m.

Diphyes dispar

Image credit: Denis Riek, WoRMS

This subtropical species (Fig. **49**) is found distributed in the Pacific Ocean, Atlantic Ocean and the Mediterranean at a depth range of 0-500 m. Nectosac of anterior nectophore of this species is cylindrical basally but with a narrow caecal extension ending close to the tip of the nectophore. Hydroecium extends to one-half the height of its nectophore. The dorsal ostial tooth is y larger than the lateral teeth. The mouth plate is not divided. The posterior nectophore has a more or less similar arrangement of ostial teeth. Lateral teeth and baso-lateral margins are not serrated. Phyllocyst tapers towards its apex.

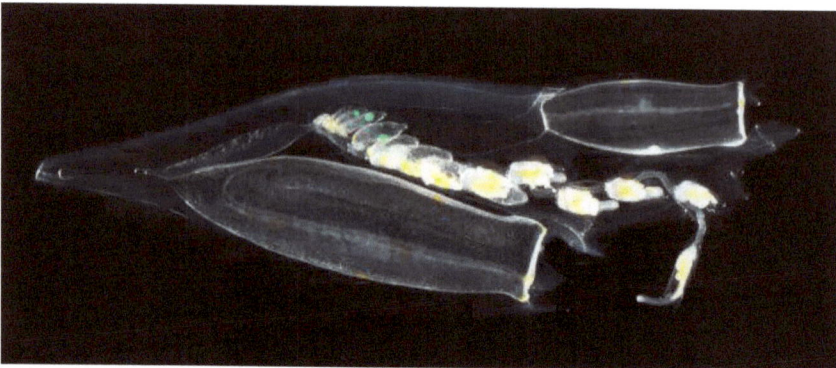

Fig. (49). *Diphyes dispar.*

Bioluminescence: This species is not truly bioluminescent but it bears fluorescent spots along with the stem of its siphosome or on the tentilla themselves. However, bioluminescence has been observed in the gastrozooids, bracts and nectophores [17].

Erenna spp.

Image credit: The Oceanography Society (Reproduced with permission)

The luminescent species of this genus *viz. Erenna laciniata* (Fig. **50**), *Erenna richardi, Erenna cornuta, Erenna sirenia and Erenna insidiator* occur at aq depth range of 1600-2300 m in Europe and Northern America and Southestern Asia. Nectophores are dorso-ventrally flattened with tapering axial wings. Apico- and infra-lateral ridges form upper and lower margins of lateral surface respectively. Lateral radial canals are straight, thickened on apico-lateral margins of nectosac. Bracts are of two types which are with patches of epidermal cells (including nematocysts) on dorsal swelling at distal extremity. Tentillum is large, with hypertrophied, uncoiled cnidoband, and rigid terminal process. Gastrozooid has large swollen basigaster.

Fig. (50). *Erenna laciniata.*

Bioluminescence: The side branches (tentilla) tentacles (Fig. **51**) of *Erenna* spp. have been reported to emit pulses of red light and these structures serve as wriggling lures to attract their prey.

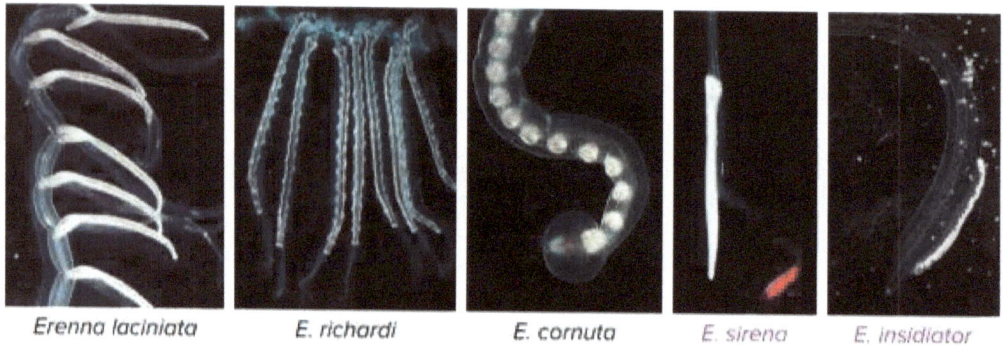

Erenna laciniata   E. richardi   E. cornuta   E. sirena   E. insidiator

Fig. (51). Luminescent tentacles of *Erenna* spp.

Each tentillum consists of a large cnidoband (several stinging cells) which is attached to a central stalk. This transparent stalk terminates in a bulb containing white spots called Bocelli. When ruptured these spots produced luminescence, indicating that they are also photophores filled with $Ca2\text{þ}$-regulated photoproteins. Unlike typical cnidarian photocytes, the terminal photophores of these siphonophores did not flash readily on direct stimulation. It is also reported that the photophores of young tentilla of these species contain only bioluminescent tissue, and when they mature, they get surrounded by the red fluorescent material. This substance actually produced a multimodal fluorescence emission varying from yellow to red (583, 620, and 680 nm) . These mature tentilla have also been reported to display a unique rhythmic flicking behavior [76].

Forskalia edwardsi

Image credit: Schuchert, Peter, WoRMS

Forskalia edwardsi is a shallowater species (Fig. **52**) living at depths of 0-200 m in the tropical and subtropical Pacific Ocean and Atlantic Ocean. Anterior portion of the colony consists of a small float and a cylindrical region of numerous nectophores which are arranged in multiple rows. A stem of several metres long follows the nectophore swhich are oddly-shaped, boxy, and transparent.

Fig. (52). *Forskalia edwardsi.*

Bioluminescence: When disturbed, the palpons (the reddish stinging batons) of this luminescent species have been reported to eject a plum or orange coloured drop of bioluminescent liquid from their tips [77].

Frillagalma vityazi

Image credit: The Oceanography Society (Reproduced with permission)

The luminescent species, *Frillagalma vityazi* (Fig. **53**) has been recorded from the surface to 2000 m, together with a single record from the 3330-3910 m depth range in Indian and North Atlantic Oceans. It is said to be common off British Columbia (NE Pacific) (Pugh,1998). The main features of the small nectophores of this species are that all the ridges are flared and frilled; and there are two pairs of short ridges arising from the apico-laterals of these nectophores. Courses of the lateral radial canals on the nectosac are simple and unlooped.

Bioluminescence: The patches of ectodermal cells found on both the nectophores and bracts have been found to be the sites of bioluminescence in this species. The bioluminescence of this species is either blue or green, and it is worthy of mention here that the nectophores of some specimens of this species have produced bioluminescence of one colour and the bracts produced another colour. Further, in this species, the peak of a broad unimodal spectrum moved from one end of the range to the other [57].

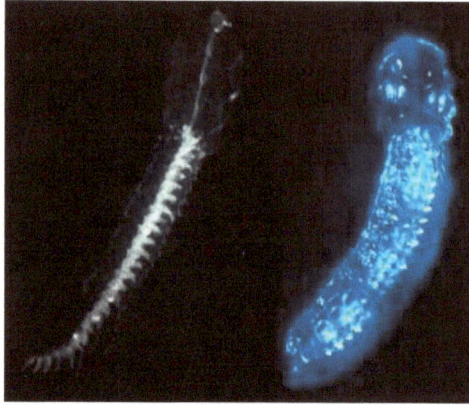

Fig. (53). Bioluminescent *Frillagalma vityazi*.

Hippopodius hippopus

Image credit: Catriona Munro, Stefan Siebert, Felipe Zapata, Mark Howison, Alejandro Damian-Serrano, Samuel H. Church, Freya E. Goetz, Philip R. Pugh, Steven H. D. Haddock, Casey W. Dunn, Wikimedia.

The shallow water species, *Hippopodius hippopus* (Fig. **54**) is found at depths of 0- 200 m in the tropical and subtropical Pacific Ocean, Atlantic Ocean and the Mediterranean. It is a bullet-shaped, very delicate colony in which 10 horseshoe-shaped nectophores fit together alternating on opposite sides around the yellowish stem. Individual nectophores are about 1cm in diameter and 3-4mm thick.

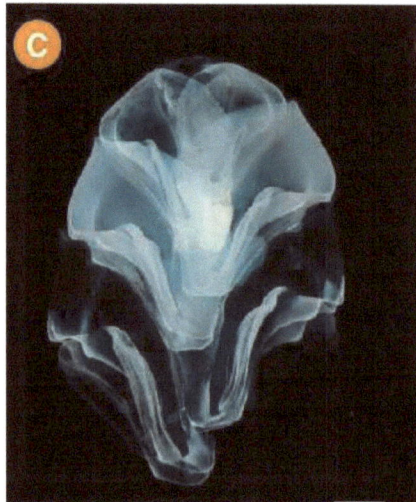

Fig. (54). *Hippopodius hippopus*.

Bioluminescence: The nectophores of this species are brilliantly bioluminescent in life and become cloudy whitish as they die [77]. In the species of *Hippopodius,* blue light is produced as waves which spread over the surface of the nectophores. In these species light is produced within the ectoderm and is said to be propagated by electrical impulses in this tissue [78].

Maresearsia praeclara

Image not available

It is a neritic and oceanic species found distributed in tropical, subtropical and temperate waters of Pacific, Atlantic Oceans and the Antarctic. The two nectophore of this luminescent species which are with a functional flask-shaped nectosac, fit together to form a ball-like structure.

Muggiaea sp.

Image credit: Anne Wesche, Wikimedia

The members of the genus *Muggiaea* (Fig. **55**) are colonial siphonophores and are often with two nectophores (swimming bells) arranged one behind the other. But in the luminescent species of the genus *Muggiaea*, the posterior nectophore is absent and the anterior one has a characteristic complete dorsal ridge. The somatocyst (extension of the gastrovascular system) is found to be very close to the wall of the nectosac (central cavity with muscular walls).

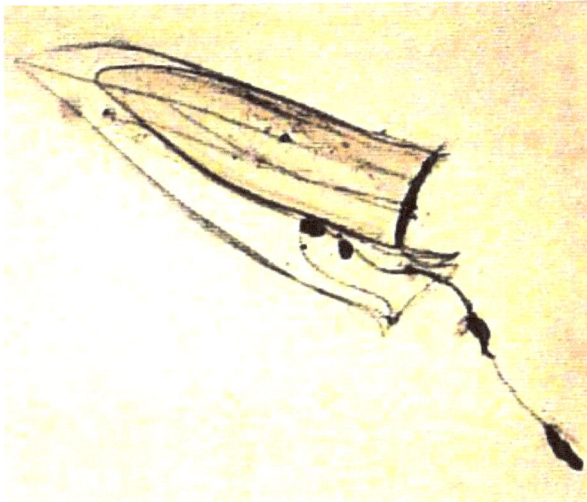

Fig. (55). *Muggiaea* sp.

Nanomia bijuga

Image credit: Neil McDaniel, WoRMS

Nanomia bijuga (Fig. **56**) has been recorded in depths up to 1800 m in Pacific, Atlantic, and Indian Oceans; warm and temperate waters. It is one of the most common physonect siphonophores off the west coast of the United States. Its pneumatophore (gas-filled float) has a red-pigmented cap. Some nectophores (swimming bells) of this species are inflated and cube-shaped. While the length of the total chain is up to 45 cm, its pneumatophore and nectophore have a size range of 3-6 mm and 2-6 mm respectively.

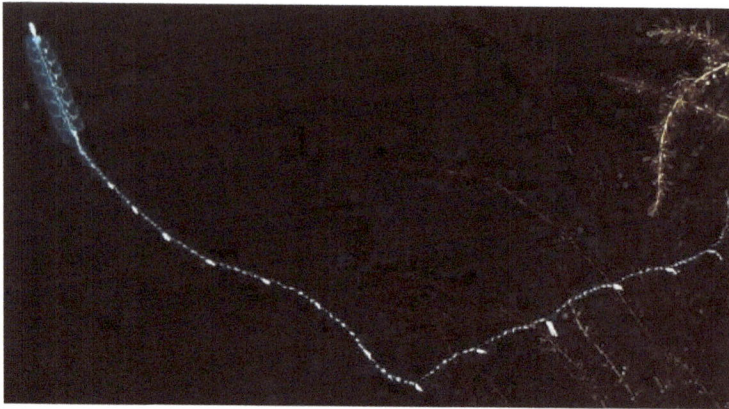

Fig. (56). *Nanomia bijuga.*

Bioluminescence: The spatial and temporal pattern of mechanically stimulated bioluminescence which was recorded *in situ* 32 times, both as controlled strikes and during transects, was sufficiently unique to make it readily identifiable. In this species, bioluminescence was found originated from both the nectosome and the siphosome . While nectosome emission was a steady glow following stimulation, lasting approximately 4 s, siphosome emission was a steady glow of about 2 s duration, followed by a brief period of scintillation in which the individual bracts yielded 200 to 300 ms flashes. Further, bracts, even after their separation from the siphosome, continued to luminesce [60].

Nanomia cara

Image credit: M. Youngbluth (NOAA), Wikimedia

Though the species, *Nanomia cara* (Fig. **57**) is well adapted to deep waters, it appears occasionally in large numbers in the surface. It is a common species of the entire North Atlantic Ocean. Colony of this species can attain a maximum length of several meters. Following the pneumatophore, a large number of nectophores (swimming bells) form a 5-20 cm long thickening of the colony. These swimming bells are individual animals providing propulsion to the entire colony.

Fig. (57). *Nanomia cara.*

Bioluminescence: In the species of *Nanomia*, patches of luminescent cells are seen in the nectophores in which these cells form two oblong patches located symmetrically on either side of the exumbrella near the margin. Further its bracts also possess two symmetrical photophores on the upper side near the tip. Freeman (1987) reported that the base of the tentacle of its larva has the ability to produce light for a few days. Further, paired bilaterally symmetrical bioluminescent organs also develop on the nectophores and the bracts of the adult colony. In both the larva and the adult of this species, the bioluminescence is mediated by a calcium specific photoprotein. However, the photocytes of these organs lack a green fluorescent protein.

Nectadamas diomedeae

Image not available

This luminescent species has been recorded from Atlantic and Pacific Oceanic areas *viz.* Brazil, Canada and USA. It has a singular, rhomboidal definitive nectophore, which bears a distinctive pattern of ridges. Somatocyst canal system is complexly divided. Nectophore gets distorted so that the ostium of the nectosac comes to open on its right-hand side. A deep, obliquely slanted and pocket-shaped, hydroecium is present and it has a narrow opening. Bract is more or less triangular in shape. Tentillum of the tentacle is characteristically shaped, and it does not possess a terminal filament. Cnidoband consists of a large, sub-terminal, hemispherical swelling and a terminal cap.

Nectopyramis natans

Image not available

This luminescent species has been recorded from Antarctic, Atlantic and Pacific Oceanic areas *viz.* South Orkney, Brazil, Canada and USA. It has an elongate, bow-shaped definitive nectophore with a pointed apex and truncated base. There are seven longitudinal ridges which are arranged characteristically. A simple somatocyst without branches is present. Larval nectophore has a nectosac and 4 radial canals which are arising separately from the simple somatocyst. Its bow-shaped eudoxid bract is with 5 longitudinal ridges.

Physalia physalis

Image credit: Biusch, Wikimedia

Physalia physalis (Fig. **58**) otherwise called as "The Portuguese man-of-war" lives in warm tropical and subtropical water such as the Florida Keys and Atlantic coast, the Gulf Stream, the Gulf of Mexico, the Indian Ocean, the Caribbean Sea, and other warm areas of the Atlantic and Pacific oceans. It is also very common in the warm waters of the Sargasso Sea. It is a colony consisting of four types of polyps: a pneumatophore, or float; dactylozooids, or tentacles; gastrozooids, or feeding zooids; and gonozooids. Cnidocytes (stinging cells) are present in the tentacles.

Fig. (58). *Physalia physalis.*

Bioluminescence: Its red fluorescence reported is due to the combination of biliprotein and artifact [17].

Praya dubia

Image credit: Catriona Munro, Stefan Siebert, Felipe Zapata, Mark Howison, Alejandro Damian-Serrano, Samuel H. Church, Freya E.Goetz, Philip R. Pugh, Steven H.D.Haddock, Casey W.Dunn, Wikimedia.

Praya dubia is a subtropical species (Fig. **59**) which is otherwise called as "giant siphonophore" or "living drifting net" which occurs at a depth range of 700 - 1000 m in Atlantic Ocean, Pacific Ocean and the Antarctic. With a body length of about 50 m, *Praya dubia* is the second-longest sea organism, after the bootlace worm which measures 55 m. It has large paired transparent swimming bells at the head which is followed by a long stem made up of units called cormidia.

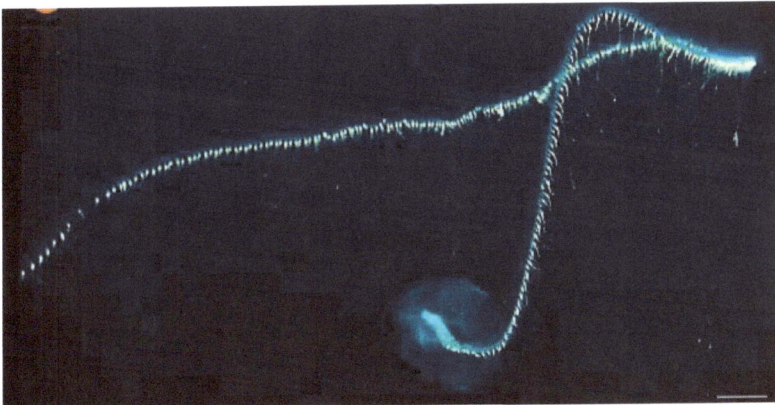

Fig, (59). *Praya dubia.*

Bioluminescence: It produces a beautiful blue bioluminescent glow to attract its prey [79].

Rhizophysa eysenhardti

Image credit: Catriona Munro, Stefan Siebert, Felipe Zapata, Mark Howison, Alejandro Damian-Serrano, Samuel H. Church, Freya E.Goetz, Philip R. Pugh, Steven H.D.Haddock, Casey W.Dunn, Wikimedia

Rhizophysa eysenhardti (Fig. **60**) is a tropical and subtropical species and is found in Eastern Pacific areas *viz.* Mexico, Costa Rica, USA and Canada. Mature animals of this species are pale pink in colour. Hypocystic villi are seen at the base of pneumatophore. Tentacles are provided with filiform tentilla.

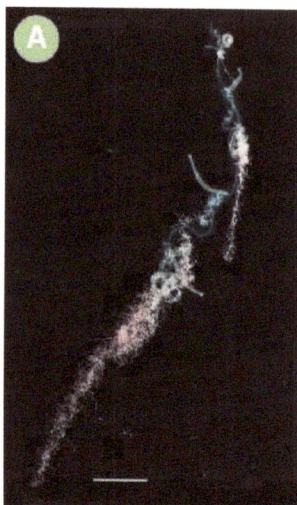

Fig. (60). *Rhizophysa eysenhardti.*

Bioluminescence: This species is not bioluminescent. But its brightly colored and motile tentilla are fluorescent [17].

Rosacea plicata

Image not available

This species has Cosmopolitan distribution from Bering Sea to the Antarctic and its depth range is 300 - 500 m. Nectophores of this species are smooth and are more or less rounded-rectangular in outline. They have thick, stiff jelly and a hollowed cavity in the posterior portion of each nectophore. These nectophores are transparent with a yellow to the orange stem. The size of the colony is 15-60mm long and, 10-15mm wide.

Bioluminescence: The larva of this species was luminescent and its emission wavelength maximum was found to be 488 nm [23].The gastrozooids of this species are fluorescent; and their bracts and nectophores (swimming bells) are bioluminescent [17].

Stephanomia amphytridis (= *Halistemma amphytridis*)

Image credit: YouTube

Stephanomia amphytridis (Fig. **61**) is fairly a deep-sea species found in Pacific Ocean areas such as USA and Canada. It grows to a length of 10 metres or more, although it is only 15-20 cm in diameter. It is a colony of specialised individuals called "zooids". Its nectophores are larger with two vertical-lateral ridges. Lateral radial canals of the nectosac are more extensively looped. . Siphosomal zooids of this species include gastrozooids, palpons, bracts and gonophores . Its mature tentillum (on the gastrozooid tentacle) is unicornuate.

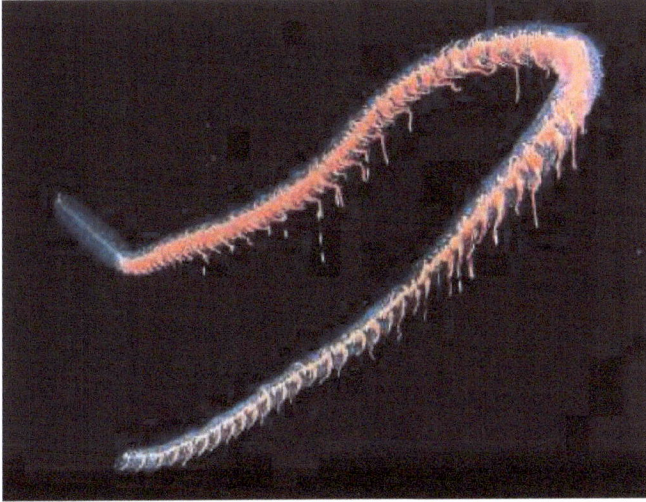

Fig. (61). *Stephanomia amphytridis.*

Bioluminescence: In this species, the bioluminescence occurs in patches along the clear nectophores (swimming bells) and bracts (leaf-shaped shields present in the long stem.) [80].

Sulculeolaria sp.

Image not available

In this luminescent species, anterior nectophore is boot-shaped, albeit it is a small one. Posterior nectophore is more cylindrical. Nectophores which are of a stiff gelatinous consistency, lack longitudinal ridges . They are conical and closed at the anterior end, with the posterior opening surrounded by gelatinous 'ostial teeth'. Both anterior and posterior nectophores are about 20mm long.

Vogtia glabra

Image not available

This subtropical, luminescent species is found in the Atlantic, Pacific Oceans and the Mediterranean. It is very common in Brazil, Canada, and USA. It is more broadly horse-shoe shaped with 2 dorso-lateral protuberances.

Bioluminescence: In this species, the blue light is produced in waves and they spread diffusely over the surface of the nectophores. The spectral emission maxima were found to be 470 nm and 455 nm [78].

Vogtia serrata

Image credit: Fisheries and Oceans Canada, Moira Galbraith, WoRMS

Vogtia serrata (Fig. **62**) is found distributed in the Atlantic, Pacific Oceans, Mediterranean and Antarctic. It is believed to be common in Brazil, Canada, USA and Antarctica. It has a roughly triangular nectophore with distinctive ridges. However there are no spines or protuberances.

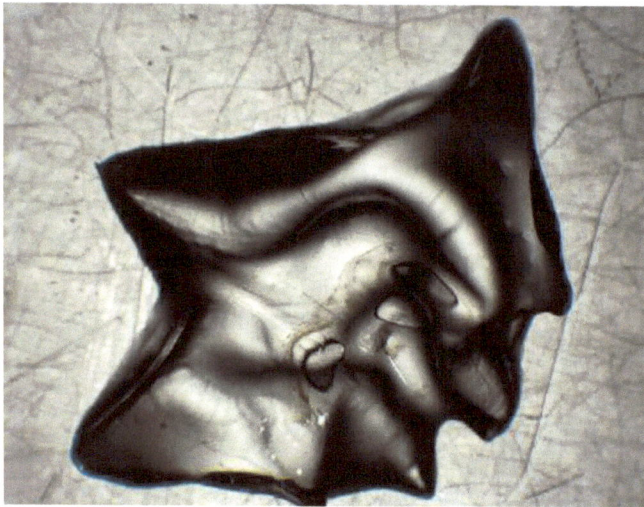

Fig. (62). *Vogtia serrata.*

Vogtia spinosa

Image credit: Fisheries and Oceans Canada, Moira Galbraith, WoRMS

Vogtia spinosa (Fig. **63**) is found in the Atlantic, Pacific Oceans and the Mediterranean and is believed to be common in Brazil, Canada and USA. It has a roughly quadrangular nectophore with numerous gelatinous teeth on the ridges.

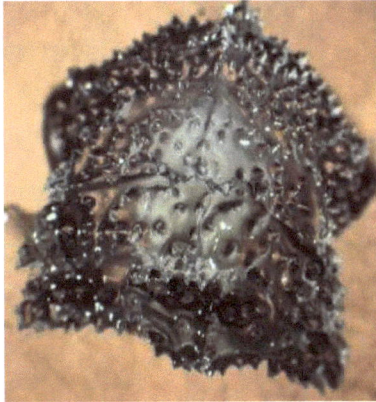

Fig. (63). *Vogtia spinosa.*

Bioluminescence: The nectophores of this species are highly luminescent in life, even when they are separated from the colony [77]. This species produced blue light in waves which spread diffusely over the surface of the nectophores [78]. Mechanical disturbance evoked luminescence in this species and it produced faint blue fluorescence with short-wavelength [57].

BIOLUMINESCENT MARINE SCYPHOZOAN MEDUSAE

The Scyphozoa which are exclusively marine are commonly known as the true jellyfish or "true jellies". There are more than 200 jellyfish species that are found distributed throughout the world's oceans, from the surface to great depths. Many species of jellyfish are known to prefer coastal waters. These jellyfish have radial symmetry and are diploblastic, *i.e.* their body wall consists of an outer epidermis (ectoderm) and an inner gastrodermis (endoderm), which are separated by mesoglea. They have nematocysts, characteristic of the phylum Cnidaria. These jellyfish undergo alternation of generations, with the medusa form being dominant. Unlike the hydromedusae, these scyphomedusae are usually large and are often highly pigmented. Most of the large and colorful jellyfish are found in coastal waters throughout the world. They normally range from 2 to 40 cm in diameter, and the largest species, *Cyanea capillata* has been reported to reach a maximum diameter of 2 m.

Bioluminescence In Marine Scyphozoans

Among the marine scyphozoan medusae 12 species (Table **3**) have been reported to be luminescent [13]. Unlike the hydrozoans, bioluminescent scyphozoan medusae have no green fluorescent protein (GFP) and some utilise photoproteins where others possess the classic luciferin and luciferase biochemistry (Haddock and Case, 1999). The mean emission max for all scyphomedusa species was found to be 473.8 nm. There was no significant difference in the mean wavelength between scyphomedusae (474.0 nm) and hydromedusae (473.7 nm) [57].

Table 3. Luminescent marine scyphozoan medusae [13]

Class	Order	Family	Species
Scyphozoa	Coronatae	Atollidae	*Atolla parva, A. vanhoeffeni, A. wyville*
-	-	Nausithoidae	*Nausithoë atlantica, N. globifera*
-	-	Paraphyllinidae	*Paraphyllina ransoni*
-	-	Periphyllidae	*Periphylla periphylla, Periphyllopsis braueri*
-	-	Ulmaridae	*Phacellophora camtschatica, Poralia rufescens*
-	Semaeostomeae	Pelagiidae	*Chrysaora hysoscella, Pelagia noctiluca*

Emission Maxima in Luminescent Marine Scyphozoan Medusae: The emission maxima (λmax) of the different species of marine hydrozoan medusae and siphonophores have been reported to vary from 442 to 491 nm (Table 4).

Table 4. Emission maxima (λmax) of luminous scyphozoan medusa.

Species	λmax	Ref
Atolla parva	442, 445 nm	[13]
-	468nm	[57]
Atolla vanhoeffeni	469nm	[81]
Atolla wyvillei	462nm	[42]
-	470nm	[10]
Chrysaora hysoscella	478 nm	[13]
Nausithoe atlantica	480 nm	[57]
Nausithoe globifera	494 nm	[57]
Paraphyllina ransoni	465nm	[57]
Pelagia noctiluca	475nm	[10]
Periphylla periphylla	465nm	[81]
-	463nm	[42]
-	470-480nm	[82]
-	470,475,480nm	[10]
Periphyllopsis braueri	473nm	[82]
Phacellophora camtschatica	491nm	[57]
Poralia rufescens	468 nm	[13]

Luminescent Marine Scyphozoan Medusae

Atolla parva

Image credit: WoRMS

Atolla parva (Fig. **64**) is a cosmopolitan species found in deep waters with a depth range of 350 - 2500 m. It is a small medusa and is up to 3 cm wide. Basal attachment of stomach is clover-shaped. Marginal septa are straight and are covered by coronal muscle;. There are 20 or 26 tentacles.

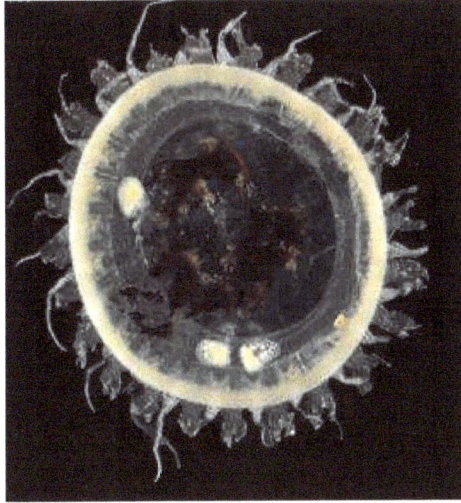

Fig. (64). *Atolla parva.*

Bioluminescence: In this species, mechanical stimulation yielded blue luminescence is associated with greater release of luminous mucus [83, 60]. Further its bioluminescence appeared in the form of one or two discrete thin streamers of light from one point on the lateral margin and such flashes lasted for several tens of seconds as a coherent luminous thread [82].

Atolla vanhoeffeni

Image credit: Fisheries and Oceans Canada, WoRMS

Atolla vanhoeffeni (Fig. **65**) occurs at a depth range, 0 - 1000 m in the Atlantic and Pacific Oceans especially Canada, Ireland and USA. Its body shape is like a flying saucer, with a conspicuous circular furrow separating the sculpted outer margin from the broad shallow central dome. There are 20 marginal tentacles, in alternation with as many statocysts. Body of the medusa is quite transparent and colorless. Its stomach is pigmented dark red. A distinct pair of black pigment spots are found on the end of each arm of the stomach proximal to the gonads. The maximum diameter of the bell is 35mm.

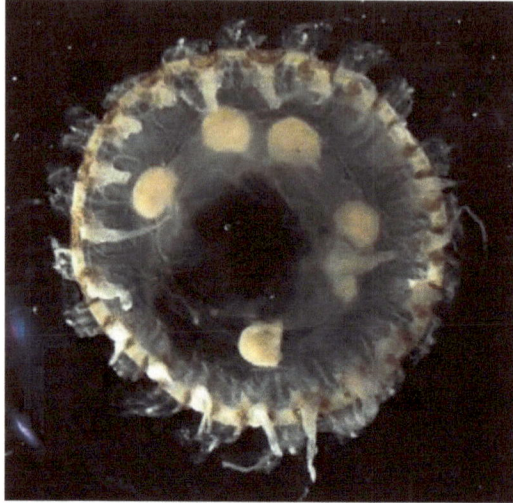

Fig. (65). *Atolla vanhoeffeni.*

Bioluminescence: This species shows luminescence down the lappets, at the tentacle bases and in the coronal groove region [82]. The gut of this species emitted blue-green light the wavelength of which was lower than the average [84].

Atolla wyvillei

Image credit: NOAA Ocean Explorer from USA, Wikimedia

Atolla wyvillei (Fig. **66**) is commonly called as "alarm jellyfish" and it occurs in deep oceans (1000-4000m) throughout the world, from the Arctic to the Antarctic. This species is also common in the North Atlantic and Pacific, the Gulf of Mexico and waters around New Zealand. The body of this species is shaped like a flying saucer and it reaches a maximum width of 20 cm. It has 22 marginal tentacles around the rim of the body and one hypertrophied tentacle, which may be 36 times longer than its bell's diameter. The stomach is broad, pigmented dark reddish-brown or almost black.

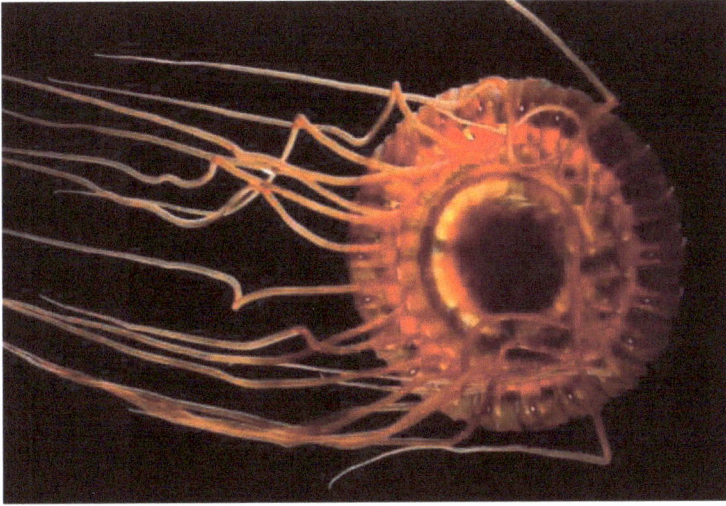

Fig. (66). *Atolla wyvillei.*

Bioluminescence: Its blue luminescence which is due to coelenterate-type (luciferin) and mechanical stimulation is distributed widely over the ex-umbrella and subumbrella. This light is said to be particularly concentrated at the bases of the marginal lappets, tentacles, ovaries and nearby coronal groove [83, 85]. It is also reported that an apparent *in situ* secretion was from the tentacles of this species [82]. Further, it is also known that almost all the species of the genus Atolla live in deep oceanic areas where sunlight never reaches and they rely on their bioluminescence [86]. Among these species, Atolla wyvillei alone uses the light in a defensive strategy.

Chrysaora hysoscella

Image credit: Ernst Haeckel, Wikimedia

Chrysaora hysoscella (Fig. **67**) is commonly called as "compass jellyfish" and it is a coastal species found distributed in the northeast Atlantic Ocean, particularly in the Celtic, Irish, and North Seas. This species has also bee reported from the Mediterranean Sea and coastal South Africa. It has a thickened bell (manubrium) which may grow up to 30 cm in diameter. Edges of the bell are developed into 32 lobes. There are 24 marginal tentacles which are arranged in eight groups of three that alternate with eight sensory organs. These marginal tentacles are conical in shape with a flattened thicker base and are also covered with clusters of stinging cells (nematocysts). It has a long and slender manubrium that leads onto 4 oral arms. This medusa is yellowish white in color with a distinctive brown pattern like the radii of a compass.

Fig. (67). *Chrysaora hysoscella.*

Bioluminescence: While light emission is intracellular in most of the medusae, it produces a luminescent slime [23].

Nausithoe atlantica

Image credit: F.S.Russell

Nausithoe atlantica (Fig. **68**) is found distributed in the subtropical Atlantic and Pacific oceanic areas at a depth range of 0-800 m. The umbrella of this luminescent species is 3.5 cm wide and is uniformly coloured in chocolate red. Stomach walls are y densely coloured. The underside of umbrella is dark. Exumbrella is smooth. Gonads are coalescent in the interradii. Tentacular pedalia are longer but less prominent than the rhopalar.

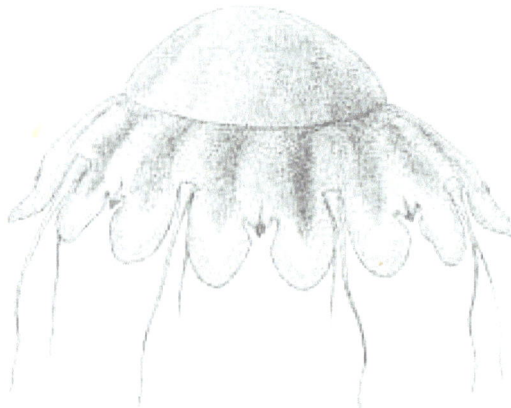

Fig. (68). *Nausithoe atlantica.*

Nausithoe globifera

Image credit: F.S.Russell

Nausithoe globifera (Fig. **69**) is deep water species occurring at a depth range of 1000 - 1333 m in Eastern Atlantic aeas such as from Greenland to Iceland, east to Bay of Biscay and south to northern Mauritania. This luminescent medusa is 1.7 cm wide. Its central disk is high, arched, solid, and is covered with nettle-spots. Pedalia is not prominent. Lappets are broad and rounded and equidistant. Gonads are large, quadrangular and are in pairs that close together. Stomach is brownish or black and gonads are light brownish, yellowish or reddish.

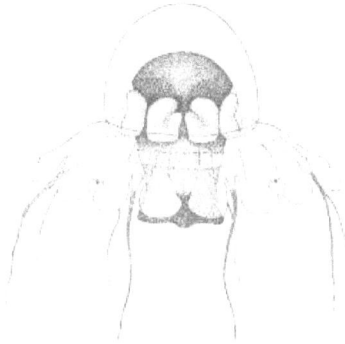

Fig. (69). *Nausithoe globifera.*

Nausithoe rubra

Image credit: Wikimedia

Nausithoe rubra (Fig. **70**) is a Shallowater species found at a depth of 0-200 m in S Atlantic, Indian, and Eastern Tropical Pacific oceanic areas.

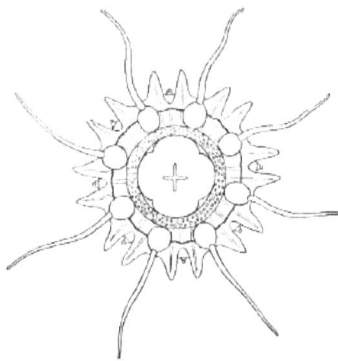

Fig. (70). *Nausithoe rubra.*

It is a small, less hemispherical medusa with a smooth surface and evenly circular coronal groove. Maximum disc diameter is 35 mm and height is half the width. There are 16 marginal lappets with alternating 8 tentacles and 8 rhopalia. Tentacle clefts are deeper than rhopalial ones. Circular muscle is more than half lappet-length in radial extent. There are 8 gonads that are contiguous in pairs. Its endodermis is chocolate-red throughout; mesoglea is lightly yellowish-brown throughout; and stomach walls are deeply colored.

Bioluminescence: Its bioluminescence is very much similar to that of *Atolla* spp., except that the dome above the coronal groove was luminescent. When the lappet margins were stimulated, a flash was initiated and this propagated radially inwards until it reached the coronal groove [82].

Paraphyllina intermedia

Image credit: Mayor, Alfred Goldsborough, Wikimedia

Paraphyllina intermedia (Fig. **71**) is found distributed in the tropical;Indo-Pacific from Indian Ocean to the west coast of USA. This medusa is 1.5 cm wide, 0.8 cm high and flatly rounded, without a pointed apex. Marginal lappets are oval and are bluntly pointed. Its 16 pedalia are rectangular with rounded angles. The four rhopalar pedalia are half as wide as the 12 tentacular pedalia. Gonads are in four pairs and are of bean-shaped or egg-shaped sacs. Ocelli are with the lens. The stomach has red-brown coloration in interradial parts.

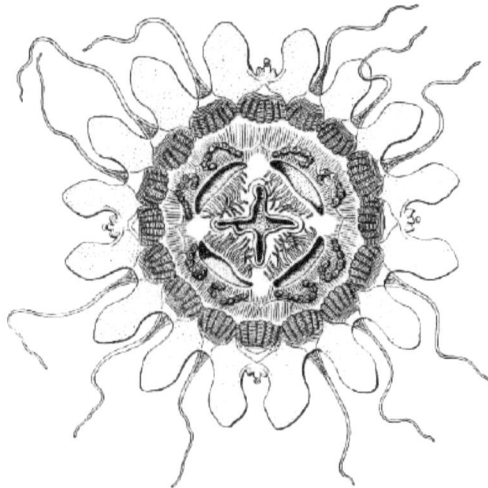

Fig. (71). *Paraphyllina intermedia.*

Bioluminescence:. The luminescence of this species was clearly composed of bright point sources than that of *Atolla* spp. or *Nausithoe rubra*. In this medusa, a mechanical stimulus in the coronal groove region produced bioluminescence down the outer margins of each lappet and the junction between adjacent lappets produced a V-shaped luminescence pattern. This luminescence spread as a wave both radially and over the exumbrellar surface and a strong stimulus finally resulted in a series of such waves [82].

Paraphyllina ransoni

Image credit: Morandini, André, WoRMS

Paraphyllina ransoni (Fig. **72**) is a deepwater species found at a depth of 1500 m in the Pacific, Atlantic and Mediterranean areas such as Liberia, USA and Canada. The Bell of this medusa is up to 3.5 cm wide. Umbrella has a dome-shaped apex. Rhopalar pedalia are slightly narrower than the tentacular ones. Marginal lappets are evenly rounded. There are eight adrarial gonads which are W-shaped. Ring muscle is continuous. There are no ocelli with the lens. The whole medusa has brownish-red coloration.

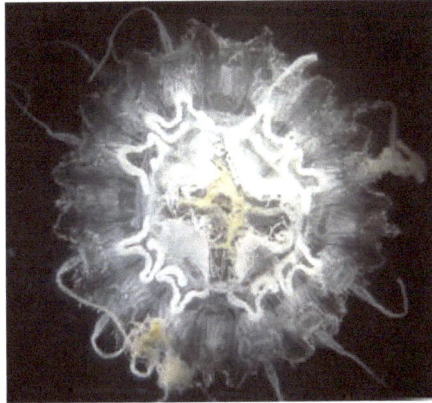

Fig. (72). *Paraphyllina ransoni.*

Bioluminescence: - In this species, luminescence was reported only from specific sites associated with the lappets and the tentacle bases [68].

Pelagia noctiluca

Image credit: Hectonichus, Wikimedia

Pelagia noctiluca (Fig. **73**) commonly called as 'Mauve Stinger' or 'Purple People Eater' is normally considered to be oceanic, but it may be washed near shore. It is found globally in tropical and temperate waters where they occur in great numbers. Bell diameter of this small medusa is only up to 90 mm. The coloration of the medusa is pinkish to purplish, with conspicuous warts on the exumbrellar surface. There are eight laterally compressed tentacles that arise from marginal notches, alternating with sensory organs called 'rhopalia'. Its four long, crenulated oral arms surround the mouth.

Fig. (73). *Pelagia noctiluca.*

Bioluminescence: This scyphozoan jellyfish is brilliantly luminescent. It uses a photoprotein and Ca2+ to produce a luminescent glow rather than a flash [72]. These medusae do not luminesce in calm water and when they are stimulated, they produce an intense blue-green light [83]. A slight mechanical stimulation at the ex-umbrella causes a spot of light at the point of touch and this light spreads out in lines, streaks or patches. Further, stronger electrical, chemical or mechanical stimuli have also been reported to cause a general glow of its exumbrella especially at the marginal lobes, oral arms or tentacles. Freshly collected specimens were also found to yield a flickering response from its subumbrellar surface. In this species, the bioluminescence is due to photocytes *i.e.* granule-containing cells and the presence of free granules in the mucus has caused luminescence and its tissue luminescence is believed to be extracellular.

Periphylla periphylla

Image credit: Edward Adrian Wilson, Wikimedia

Periphylla periphylla (Fig. **74**) is commonly called as "helmet jelly or merchant-cap" and is one of the most abundant deep-sea jellyfish. It is a circum-global, luminescent species found at depths of 0 - 2900 m .The body of the medusa is like a cone within a cone. Umbrella is up to 20 cm wide and it is usually higher than wide. It has a deep-red pigmented stomach and a sculpted rim with typically 12 whitish tentacles which are arranged in four groups of 3. Individual specimens may or may not possess an elongated projection from the stomach. Gonads are U-shaped and the coloration of stomach and subumbrella is purple or violet.

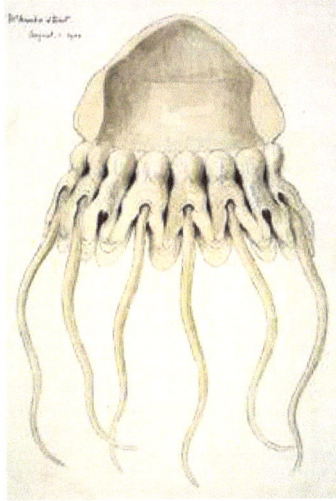

Fig. (74). *Periphylla periphylla.*

Bioluminescence: This species is known for its most varied and extensive luminescent displays. In this species, luminescent responses were found propagated round the umbrella, round each lappet and along the coronal groove, either separately or in concert. Scintillating particles were released from the lappet margins. Its ovaries were also found to emit light with an mission maximum of 470-480nm [68]. It is reported that when these medusae were stimulated with short pulses, they responded with glows as summated flashes which lasted 0.8-4 sec [61]. Further, the blue green reflectances of the gut of this species were lower than the average (~1%) [84]. Its blue luminescence was found associated with the greater release of luminous mucus [83].

Periphyllopsis braueri

Image credit: Flickr

Periphyllopsis braueri (Fig. **75**) is a circumtropical species having a depth range of 0 - 1410 m.Medusae of this species are flattened, 6 cm wide and 2.5 cm high. Its ring furrow is deep. Coronal muscle is very weak. There are eight gonads which are oval and equidistant. Its entire endodermal system chocolate-red.

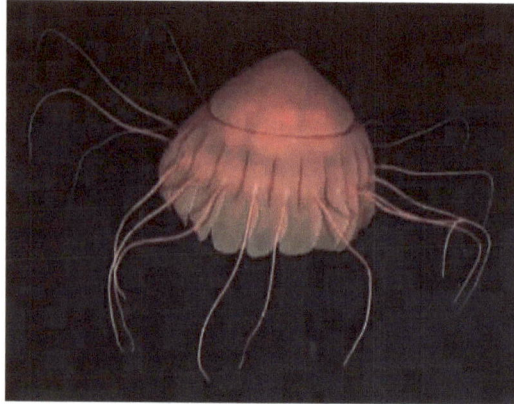

Fig. (75). *Periphyllopsis braueri.*

Bioluminescence: These medusae have been reported to respond to touch with local luminescence which consisted of a large number of rapidly pulsing sites. Further, stroking the exumbrella yielded persistent lines of luminescence. Sticky luminous material shed into the water by these medusae yielded flashing as separate particles for the periods of tens of seconds [82].

Phacellophora camtschatica

Image credit: Wikimedia

The shallow water luminescent medusa, *Phacellophora camtschatica* (Fig. **76**) is commonly found distributed in coastal environments throughout the Pacific from Alaska to southern California . It has also been reported worldwide in temperate oceans. Its maximum bell diameter is 60 cm. Its transparent margin consists of 16 large lobes that alternate with small lobes; and each lobe contains clusters of up to 25 tentacles which measure up to 6 m long. Its central yellowish gonad is surrounded by its milky whitish bell.

Fig. (76). *Phacellophora camtschatica.*

Poralia rufescens

Image credit: Neptune Canada, WoRMS

Poralia rufescens (Fig. **77**) is a cosmopolitan species found at depths ranging from 500 to 2500 m. Body of the medusa which is up to 25 cm in diameter is deep dark red throughout, with a smooth central dome over the stomach. However, this jelly fish may also be colorless with tiny red pigment granules. Umbrella margin bears numerous fine tentacles which are located between its small, rounded lappets. Radial canals are long and straight and are clearly visible on the subumbrellar surface. A total of 4-18 short, tapered oral arms are found around the mouth. Gonads form a continuous ring around the perimeter of their stomach.

Fig. (77). *Poralia rufescens.*

CONCLUSION

Though the bioluminescence is widespread in all major Cnidarian groups *viz.* hydrozoans, scyphozoans, and anthozoans, very little is known about this phenomenon in cubozoans. The importance of the luciferin involved in the chemical reaction of bioluminescence in these cnidarians *viz.* coelenterazine is well known. but the genes controlling the coelenterazine synthesis need to be identified. Further, cnidarians can survive without fluorescence or bioluminescence, but the selective advantage of the light emissions of these organisms needs to be established. More culture experiments may therefore be needed to discover the factors which affect bioluminescence in cnidarians which could contribute to overall knowledge

CHAPTER 6

Bioluminescent Ctenophores

Abstract: This chapter deals with the total luminescent fauna of the phylum, Ctenophora,; the emission maxima in observed species of ctenophores, the description of luminescent species of ctenophores, and their mechanism of bioluminescence.

Keywords: Emission maxima, Luminescent nudans, Luminescent tentaculates.

INTRODUCTION

Ctenophores or comb jellies which are probably common members of the zooplankton are presently classified under a separate phylum *viz*. Ctenophora. The latter once formed a major component of the erstwhile phylum Coelenterata. Ctenophores are exclusively marine, and they can be found in most marine habitats, from polar to tropical, inshore to offshore, and from near the surface to the very deep ocean. They are characterized by their eight rows of cilia, which are mainly used for locomotion. These organisms may be seasonally much more abundant in the spring and early summer. Almost all ctenophores are bioluminescent, with the exception of the sea-gooseberry, *Pleurobrachia*, and some benthic species. Like hydrozoans, these comb jellies use photoproteins with coelenterazine to make light. Some species secrete luminous material into the water when they are disturbed. Though there are about 100-150 species of ctenophores throughout the world's oceans, the existing information on about 50 species of bioluminescent ctenophores is scanty [87]. The present topic deals with the biology and ecology of common bioluminescent species of ctenophores.Coastal ctenophores: The common coastal ctenophore representatives of the order Cydippida are round or oblong in shape and are usually below 3 cm in diameter. These ctenophores are distinguished by their more or less solid bodies, eight radial comb rows, which are helpful in their locomotion, and two-branched tentacles for capturing small planktonic prey. The most common cydippid ctenophore species, distributed worldwide are the non-luminescent species of the genus *Pleurobrachia* andthe most common luminescent representatives of the order, Lobata include the species of the genera *Bolinopsis* and *Mnemiopsis*.

Oceanic ctenophores: Many species of ctenophores are found only far offshore near the surface, mid-water, or in the deep sea. Unlike the coastal ctenophores, oceanic ctenophores are much more fragile because they do not need to tolerate heavy wave action or the sediment load of coastal waters. Notable species of open ocean near-surface ctenophores include the Venus' girdle, *Cestum veneris* and the lesser known species include that of *Ocyropsis*.

Bioluminescence in ctenophores: Among the ctenophores, 31 species have been reported to be luminescent (Table 1). In luminescent ctenophores, both luciferin (coelenterazine) and Ca^{2+-} activated photoprotein are present [27]. One of the most important characteristics of ctenophores is their light-scattering produced by beating the eight ctenes (comb rows) of locomotory cilia. This light scattering appears as a changing rainbow of colors running down the comb rows and this is simple light diffraction or scattering of light by the moving cilia. Several species of ctenophores are also bioluminescent, but their light (usually blue or green) can only be seen in darkness. This bioluminescence is invariably caused by activating the calcium-activated proteins called photoproteins in cells called photocytes, which are normally confined to the meridional canals that underlie the eight comb rows. Species averages of maximal wavelengths for these ctenophores ranged from 440 to 506 nm. Deep-dwelling ctenophores produce light with shorter wavelengths compared to shallow-dwellers [57].

Table 1. Bioluminescent ctenophores [13].

Class	Order	Family	Species
Tentaculata	Cydippida	Haeckeliidae	*Aulacoctena acuminata, Haeckelia beehleri, H. bimaculate,H.rubra*
-	-	Bathyctenidae	*Bathyctena chuni*
-	-	Lampeidae	*Lampea lactea, L. pancerina*
-	-	Pleurobrachiidae	*Hormiphora luminosa*
-	-	Incertae Sedis	*Tizardia phosphorea*
-	-	Euplokamidae	*Euplokamis stationis*
-	-	Mertensiidae	*Charistephane fugiens, Mertensia ovum*
-	Thalassocalycida	Thalassocalycidae	*Thalassocalyce inconstans*
-	Lobata	Bathocyroidae	*Bathocyroe fosteri*
-	-	Bolinopsidae	*Bolinopsis infundibulum, B.vitrea, Mnemiopsis leidyi*
-	-	Eurhamphaeidae	*Deiopea kaloktenota, Eurhamphaea vexilligera, Kiyohimea aurita*
-	-	Leucotheidae	*Leucothea multicornis, L. pulchra*
-	-	Ocyropsidae	*Ocyropsis maculata immaculata, O.fusca*

(Table 1) cont.....

Class	Order	Family	Species
-	Cestida.	Cestidae	*Cestum veneris, Velamen parallelum*
Nuda	Beroida	Beroidae	*Beroe abyssicola, B. cucumis, B. forskalii, B.gracilis, B.ovata*

Emission maxima in luminescent ctenophores: The emission maxima (λmax) of the different species of marine hydrozoan medusae and siphonophores have been reported to vary from 443 to 550 nm (Table **2**).

Table 2. Emission maxima (λmax) of luminous ctenophores.

Species	λmax	Ref
Aulacoctena acuminata	458nm	[57]
Bathocyroe fosteri	459 to 492nm	[57]
Bathyctena chuni	492 nm	[57]
Bathyctena spp.	488,490nm	[13]
Beroe abyssicola	491nm	[57]
Beroe cucumis	489nm	[57]
Beroe forskalii	491nm	[57]
Beroe gracilis	495nm	[88]
Beroe ovata	493nm	[57]
-	494,500nm	[10]
Bolinopsis infundibulum	488nm	[57]
-	505nm	[10]
Bolinopsis vitrea	490nm	[57]
Cestum veneris	493nm	[57]
-	490 nm	[23]
Charistephane fugiens	468nm	[57]
Clytia bakeri	508nm	[10]
Clytia edwardsi	508nm	[10]
Clytia hemisphaerica	504nm	[57]
Clytia gregaria	460, 494 nm	[60]
Deiopea kaloktenota	489nm	[57]
Euplokamis stationis	467 nm	[13]
Euplokamis sp.	483nm	[57]
Eurhamphaea vexilligera	496nm	[57]
Haeckelia beehleri	500nm	[57]

(Table 2) cont.....

Haeckelia bimaculata	490nm	[57]
Haeckelia rubra	489nm	[57]
Kiyohimea aurita	491nm	[57]
Lampea lactea	469nm	[57]
-	486 nm	[23]
Lampea pancerina	473nm	[57]
Lampea sp.	470 nm	[13]
Leucothea multicornis	488nm	[57]
Leucothea pulchra	488nm	[57]
Mnemiopsis leidyi	480,485nm	[10]
Mnemiopsis sp.	488nm	[10]
Ocyropsis maculata immaculata	489nm	[57]
Ocyropsis spp.	480,485nm	[10]
Thalassocalyce inconstans	491nm	[57]
Velamen parallelum	501nm	[57]

LUMINESCENT SPECIES OF CTENOPHORES

Class: Tentaculata

Aulacoctena acuminata

Image not available

This species is found distributed in the subtropical regions *viz.* the Atlantic and Pacific Oceans, USA and Canada. The body of the animal is quadrate to somewhat flattened in tentacular plane. The tentacles which are simple without tentilla are emerging orally. The stomodeum is with dark pigmentand has both lateral diverticula arising from the meridional canals.

Bioluminescence: The bioluminescence of this species is due to its green fuorescent proteins [57].

Bathocyroe fosteri

The luminescent species, *Bathocyroe fosteri* (Fig. **1**) is found at intermediate depths (200 - 1000 m) in all the world's oceans. It is however, very common in Western Atlantic, the Mediterranean and Eastern Pacificregions such as Canada and USA. It is often found hanging motionlessly in an upright or inverted posture. It is mostly transparent with red-pigmented inner gut walls. It has short comb rows and it measures 2-4 cm across the oral lobes.

Bioluminescence: This species produce an intracellular bioluminescence which appears as flickering thin blue lines along the comb rows (Widder,2002). It produced multiple peak wavelengths of luminescence. where the peak of a broad unimodal spectrum moved from one end of the range to the other [57].

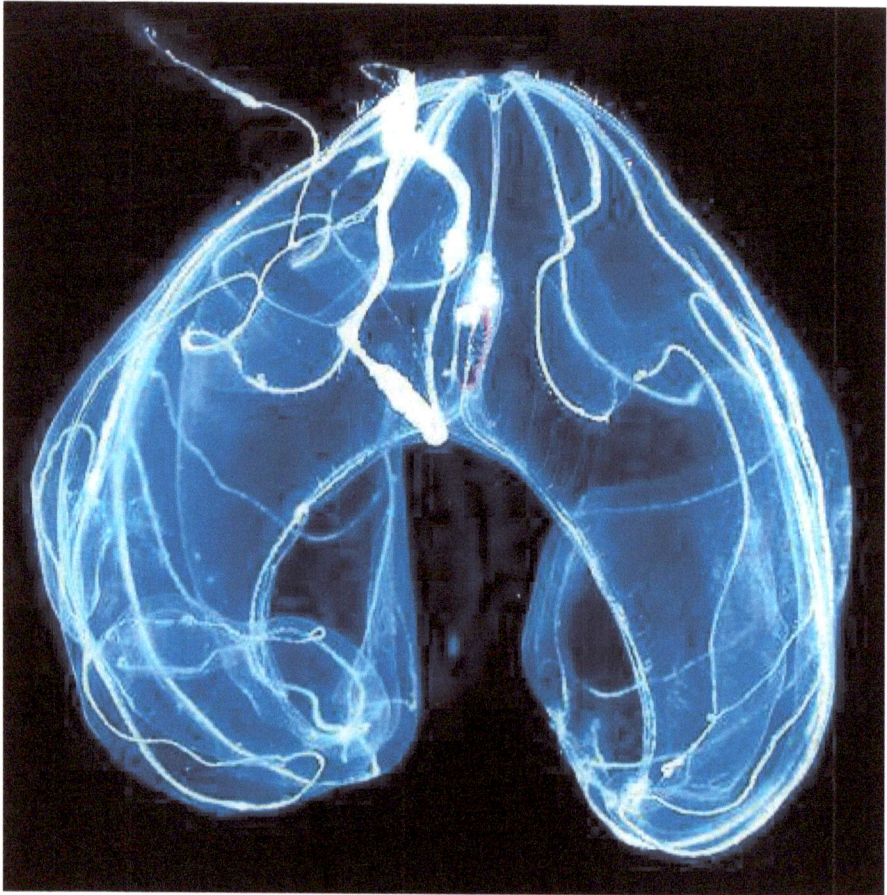

Fig. (1). *Bathocyroe fosteri.* Image credit: Marsh Youngbluth, Wikimedia

Bathyctena chuni

The deep-sea ctenophore, *Bathyctena chuni* (Fig. **2**) has cosmopolitan distribution and is found to occur at a depth range of 1000 - 3500 m. It shows a dark red color in its environment. It has its characteristic voluminous stomodæum.

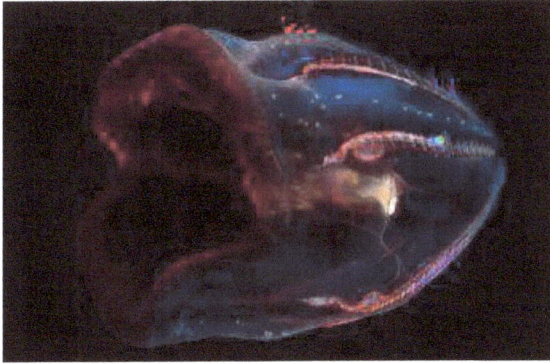

Fig. (2). *Bathyctena chuni.* Image credit: Cheryl Dybas

Bioluminescence: In such species of ctenophores, light is produced from their meridional canals and their extensions. When the animal is disturbed, it also produces ink-like secretion which luminesces as its body with the same wavelength. That is the spectra of extracellular luminescence did not differ from bioluminescence originating within the body Further, its juveniles have been reported to luminesce more brightly than the adults [57, 89].

Bolinopsis infundibulum

Bolinopsis infundibulu (Fig. **3**) is commonly known as the northern comb jelly and is found in the shallow parts (100 m) of the northern Atlantic Ocean. It is an extremely fragile species attaining a maximum length of 15 cm. Its thin gelatinous body wall is transparent, or milky white. Two short tentacles with fringed edges are present. The mouth is at one end of the body and it has two large lobes beside it, Mouth is surrounded by a ring of tentilla (little tentacles). The other end of the body is bluntly pointed. There are four long longitudinal rows and four short rows of cilia. These cilia beat in synchrony, giving the animal its iridescent appearance.

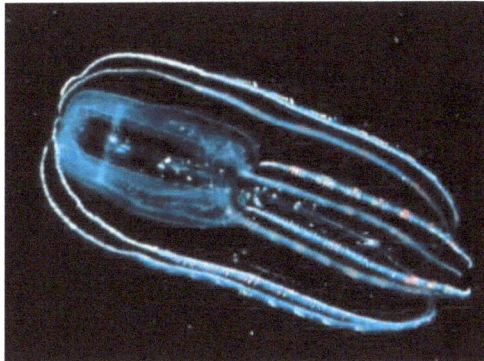

Fig. (3). *Bolinopsis infundibulu.* Image credit: NOAA Photo Gallery, Wikimedia.

Bioluminescence: In this species, the plates to which the cilia are attached are bioluminescent. It has been reported to produce the most outstanding displays, with flashes propagating around the lobes at 50 cm s^{-1} [60]. The central part of the body of the lobate ctenophores gives blue glow due to its bioluminescence while the color along their comb rows is due to light diffraction [90]. Furter, this species produced the most outstanding displays, with flashes propagating around the lobes at 50 cm s^{-1} [60].

Bolinopsis vitrea

The luminescent species, *Bolinopsis vitrea* (Fig. **4**) occurs in tropical and subtropical neritic waters of the Atlantic and Indian oceans and in the Mediterranean Sea. It is also a euryhaline species thriving in brackish waters. Though this species resembles largely *Mnemiopsis leidyi*, it is distinguished by its central origin of the lobes which form short furrows between lobes and the central portion of its body. The outer surface of the body is smooth and dark red spots may be seen inside the body. It attains a maximum length of 7 cm.

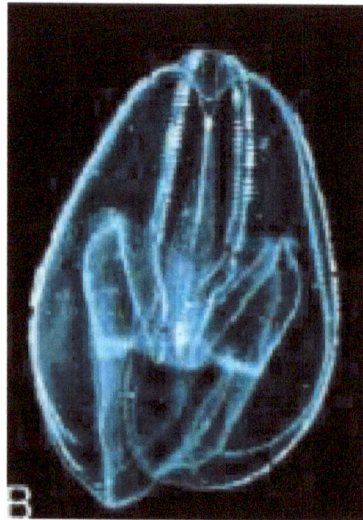

Fig. (4). *Bolinopsis vitrea.* Image credit: Prof. Tamara Shiganova (Reproduced with permission)

Cestum veneris

Cestum veneris (Fig. **5**) is commonly called as "venus girdle" and is found in the shallow waters of tropical and subtropical oceans worldwide *viz*. Atlantic and Pacific oceans, Antarctic waters, and the Mediterranean sea. It is shaped like a wing or a large transparent ribbon. It is a delicate violet in colour, but it may also be transparent with iridescent edges. While its comb rows are all on one side of the ribbon, its mouth is on the other side. Along the length of the ribbon are canals. Its unusual body form allows this animal to swim by means of muscular undulation and by using their cilia. It may grow up to a metre in total length and its width is about 5cm.

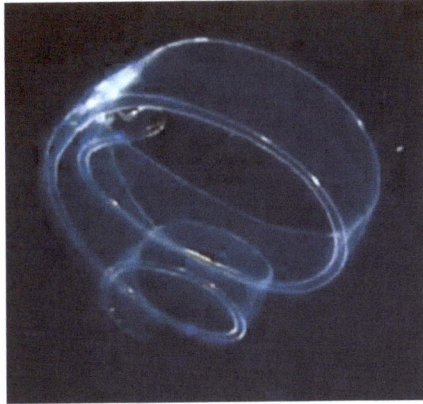

Fig. (5). *Cestum veneris.* Image credit: Dan McGanty - Wikimedia

Bioluminescence: In all cestids, brown or yellow pigment spots are believed to occur on the tips of their body. If disturbed, these animals produce brilliant bioluminescence along their meridional canals [91].

Charistephane fugiens

Image credit: The Oceanography Society (Reproduced with permission)

Bioluminescence: Like all ctenophores, this species possesses coelenterazine - based bioluminescence. It has been reported to produce its own coelenterazine for which the natural precursors are the amino acids, L-tyrosine and L-phenylalanine [92].

The luminescent species, *Charistephane fugiens* (Fig. **6**) is found in the outer continental shelf and slope at a depth of about 300 m in Western Atlantic, Eastern Pacific and the Mediterranean areas, such as Canada, Hawaii, and USA. Animals reach a size of about 2 cm. its comb rows are short and are ending oral to the midline. The body is laterally compressed and is very transparent.

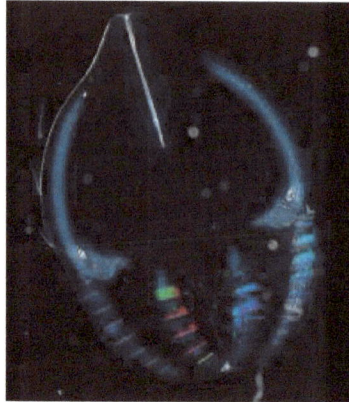

Fig. (6). *Charistephane fugiens.*

Deiopea kaloktenota

Image credit: The Oceanography Society (Reproduced with permission)

Deiopea kaloktenota (Fig. **7**) is a tropical, shallow-water species and is found distributed in Western Central Atlantic, the Mediterranean and Eastern Pacific. The morphology of this luminescent species may be simple or complex, ranging from a sac-like shape without tentacles to large lobed forms with two different kinds of tentacles.

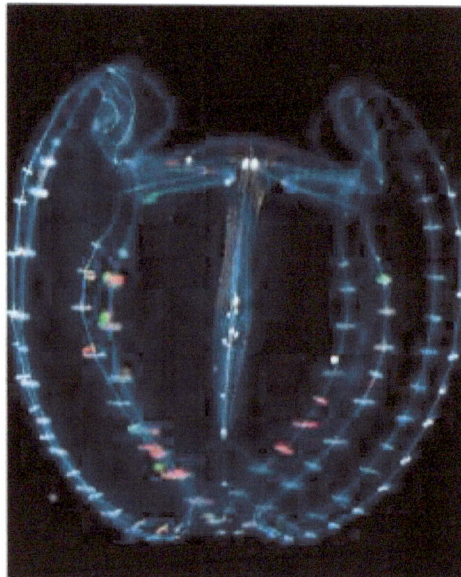

Fig. (7). *Deiopea kaloktenota.*

Dryodora glandiformis

Dryodora glandiformis (Fig. **8**) is found from boreal to polar regions of the Arctic, Atlantic Ocean, the Mediterranean and Eastern Pacific. In this luminescent species, tentacle bulbs are small, rounded, and are very nearer to the surface. Due to this, its tentacles cannot be withdrawn completely inside. Stomodaeum is deep inside a large "vestibule" at its oral end. Body is transparent and is without a greenish tint. Its size is up to 10 mm.

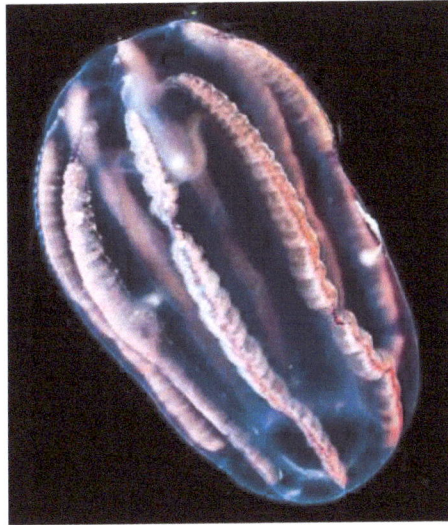

Fig. (8). *Dryodora glandiformis.* Image credit: Erling Svendsen; Norwegian Biodiversity Information Centre (Reproduced with permission)

Euplokamis dunlapae

The luminescent, shallow-water species, *Euplokamis dunlapae* (Fig. **9**) found distributed in the Pacific Ocean and in the western parts of the Atlantic Ocean. It has its characteristic branched tentacles which are coiled up to form small lumps. Its oval body which has a maximum diameter of 2 cm is transparent and is slightly colored by a red pigment. Its comb rows do not extend in full body length.

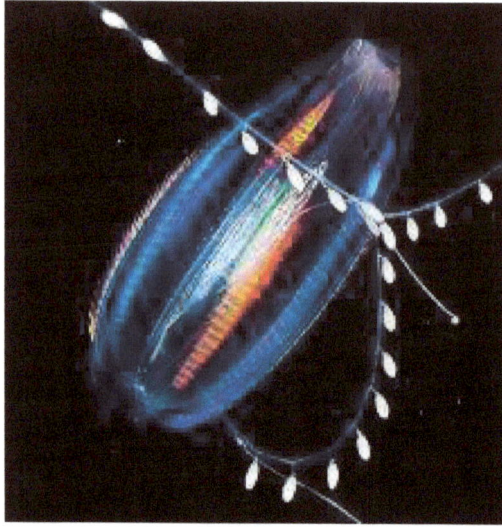

Fig. (9). *Euplokamis dunlapae* Image credit: Facebook.

Euplokamis stationis

Image credit: © S. Jamme / D.O.R.I.S. & Stéphane Jamme (Reproduced with permission)

Euplokamis stationis is fairly a deep water species (Fig. **10**) found at depths of more than 300 m in subtropical areas of the Mediterranean. Adults are more elongated with many prominent short keels which are projecting beyond the apical organ. Both adults and juveniles have transparent, bluish mesoglea with conspicuous muscle fibers. Red pigmentation is seen as rows of distinct patches on either side of the comb rows and on the tentacle bases Its coiled tentilla on the tentacles, giving the tentacle a beaded appearance and these tentilla are pinkish.

Bioluminescence: The *species of Euplokamis* are considered as the most spectacularly luminous ctenophores, as they produce not only repeated comb row flashes of high intensity but also a luminous secretion from their oral end of each comb row. The flashes of its luminous secretion could be produced independently of the comb row flashes [68].

The species of *Euplokamis* release large clouds of glowing particles in response to a mechanical disturbance. Aggregations (7-11 m^{-3}) of these ctenophores produced so much bioluminescence in response to mechanical disturbances and it was possible to read the dials and gauges in the submersible by the light of their luminescence [38].

Fig. (10). *Euplokamis stationis.*

It is also reported that when this species is disturbed, it produces ink-like secretions which luminesce as its body with the same wavelength. Interestingly its juveniles have been reported to luminesce more brightly than the adults [89].

This species has also been reported to possess green- fuorescent proteins which are responsible for its bioluminescence. In this species light is produced from the meridional canals and their extensions, but it was not limited to the regions immediately underlying the comb plates [57].

Eurhamphaea vexilligera

Image not available

It is a very shallow water species found distributed in the temperate areas of Northeast Pacific, Western Atlantic and the Mediterranean. Body of this species is with two conical processes on the aboral end, terminating in long flexible filaments. Its subtentacular comb rows extend onto these processes. The body is transparent, with red vesicles dispersed on the body. Its length is up to 10 cm.

Bioluminescence: The red vesicles dispersed on the body of this species produce a reddish-brown ink with bioluminescent properties. When the animal is disturbed, these secretions luminesce as its body with the same wavelength. Interestingly its juveniles have been reported to luminesce more brightly than the adults [89].

The spectra of extracellular luminescence associated with its luminous secretions did not differ from bioluminescence originating within the body [57].

Haeckelia beehleri

Haeckelia beehleri (Fig. **11**) is a very shallow water species found in the subtropical regions of the Eastern Pacific, Atlantic and Mediterranean. This luminescent species is about 10 mm long. Its meridional canals are equal in length and they are nearly as long as the body. Comb rows are less than one-half its body length. Stomodeal canals are seen. Long tentacle sheaths exit the body near the mouth. The body is slightly opaque with red pigments which are diffusely distributed along canals.

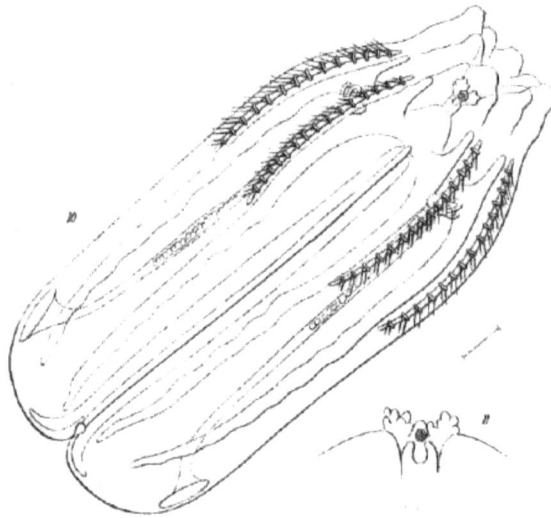

Fig. (11). *Haeckelia beehleri.* Image credit: Lindsay, Dhugal, WoRMS

Bioluminescence: Bright fluorescent granules (containing protein) are present in this species throughout its outer epithelium. Further, this species is also bioluminescent, and *in vivo* bioluminescence was at notably longer wavelengths [93].

Haeckelia bimaculata

Image not available

It is a shallow water species found distributed in tropical Western Atlantic, the Mediterranean and Eastern Pacific. Body of this luminescent species is very small, usually, 3 mm and is with rounded oral end. It has long tentacle sheaths below the mouth. Large orange-red pigment spots are seen on the canals around the base of the stomodaeum and tentacles. Small red spots are also seen along its comb rows.

Haeckelia rubra

Image credit: Lindsay, Dhugal, WoRMS

Haeckelia rubra (Fig. **12**) is a shallow water species found distributed in the tropical North Pacific and the Mediterranean. This luminescent species has a small body which is usually below 7 mm long and is with pointed oral end. Long tentacle sheaths are seen exiting the body near the mouth. There are two pairs of orange-red pigment spots on the tentacle bases.

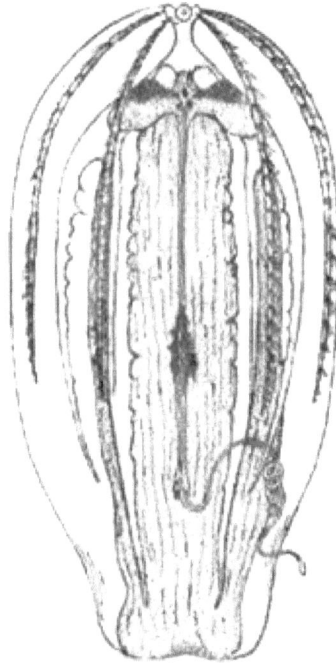

Fig. (12). *Haeckelia rubra*

Hormiphora luminosa

Image not available

The body of this species is up to 35 mm in length and is somewhat flattened at the aboral end and narrowing toward the mouth. Comb rows extend from very near the aboral pole to about three-quarters body length. Meridional canals extend beyond comb rows and branch inward toward the center of the body.

Bioluminescence: The presence of two very clear phosphorescent points in this species and these luminous organs may be responsible for its luminescence [13].

Kiyohimea aurita

Image not available

It is a tropical to subtropical, luminescent species living at depths of more than 800m in areas of Atlantic and Pacific Oceans such as the USA and Canada. Its body length may be up to 8 cm and is extremely fragile. Tentacular apparatus is absent.

Lampea lactea (= *Tinerfe lactea*)

Image not available

It is a very shallow water species found distributed in the tropical; Western Atlantic areas such as USA and Canada. It is a highly mobile, luminescent species with an extensible mouth and voluminous stomodaeum.

Lampea pancerina

Image credit: Mario Munaretto; www.Biologiamarina.org" and www.subriminigi anneri.it.(Reproduced with permission)

The tropical, luminescent species, *Lampea pancerina* (Fig. **13**) found distributed in North Pacific, Western Atlantic and the Mediterranean. The body of this species is cylindrical and is only a little narrower towards its oral end. The mouth opening is wide. Lips are dilatable to a wide sole on which this animal moves over solid surfaces. The body of the animal is transparent and it may attain a length of 7 cm.

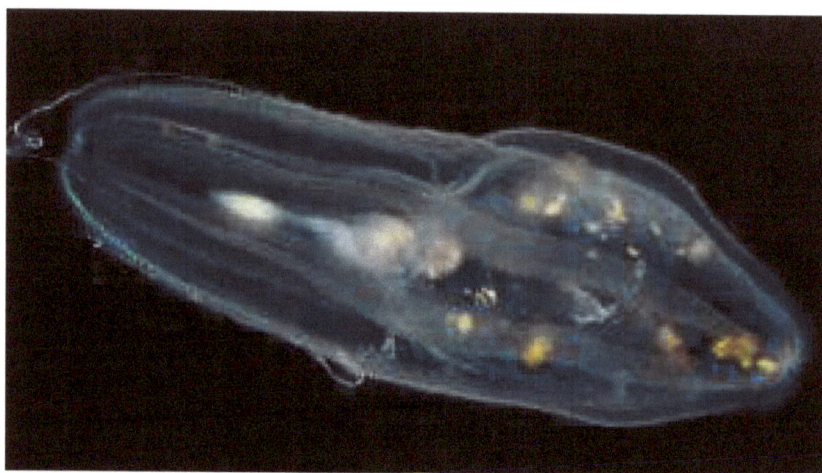

Fig. (13). *Lampea pancerina.*

Lampocteis cruentiventer

Image not available

It is commonly called as "bloody-belly comb Jelly" found at a depth of 986 m In the subtropical Northeast Pacific. This luminescent species has a red stomodaeum. Body is deeply excavated between adjacent sub-tentacular comb rows. Tentacular canals arise at infundibulum at junctions of inter-radial canals which are short branching to form adradial canals. All radial canlas join the meridional canals. Oral lobes which are of moderate size arise about half distance between mouth and aboral end of stomodaeum.

Leucothea multicornis

Image credit: Prof. Tamara Shiganova (Reproduced with permission)

The luminescent species, *Leucothea multicornis* (Fig. **14**) is widely distributed in tropical, subtropical and temperate waters in the Atlantic Ocean, Baltic, Mediterranean and Black seas, Western Indian and Southwestern Pacific Ocean . It is the largest species of the lobate ctenophores with a total length of 20 cm. It has a compressed body along the tentacular axis. Its two large lobes are about the half of the total animal length and are subdivided in two independent functional units. Both the cteneal rows are almost translucent and are arising near the aboral extremity of the body. Two opposing tentacle bulbs are seen near the mouth. Long auricles are found arising at the oral end of sub-tentacular ctene rows between mouth and base of lobes. Its color is translucent to milky white, while the pharynx and the inner portion of oral lobes are, however, yellowish.

Fig. (14). *Leucothea multicornis*

Leucothea pulchra

Image credit: WoRMS

Leucothea pulchra (Fig. **15**) is commonly called orange-tipped sea gooseberry and is one of the largest and delicate species of lobates. This subtropical, luminescent species is found distributed in Eastern Pacific areas *viz*. Canada and USA form surface to about 200 m. Body is oblong and is covered with brownish-orange papillae, which are very delicate. Its body coloration is translucent white. It attains a maximum size of 25cm.

Fig. (15). *Leucothea pulchra*

Mertensia ovum

Image courtesy: Kevin Raskoff.

The temperate to polar, luminescent species, *Mertensia ovum* (Fig. **16**) is found distributed in Black Sea, Atlantic, Pacific and Arctic Oceans from surface to 50 m depth. Body shape is oblong to spherical and is slightly flattened. Tentacle bulbs are prominent. Tentacles are provided with number of long side branches (tentilla). While the juveniles measure 2- 3 mm, adults attain a maximum size of 10 cm. The body is transparent and red/pink coloration is seen along its comb rows.

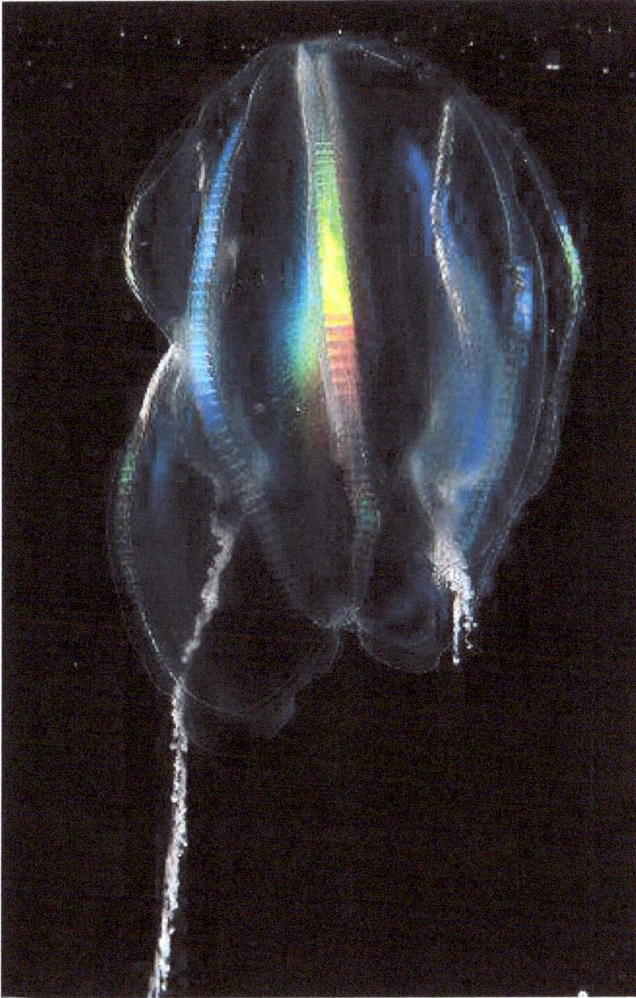

Fig. (16). *Mertensia ovum.*

Mnemiopsis leidyi

Image credit: Bruno C. Vellutini, Wikimedia

Mnemiopsis leidyi (Fig. **17**) is commonly called as sea-walnut reaching a maximum length of 10 cm. It is found distributed in the Atlantic Ocean and the Mediterranean. It is found in marine and brackish waters and is common in the intertidal, bay and nearshore, outer continental shelf and slope. Body is laterally compressed, with large lobes arising near the stomodeum. There are 4 deep, noticeable furrows that characterize the genus. It has four rows of small, but numerous, ciliated combs. It has eight meridional canals. It is usually transparent or slightly milky, translucent.

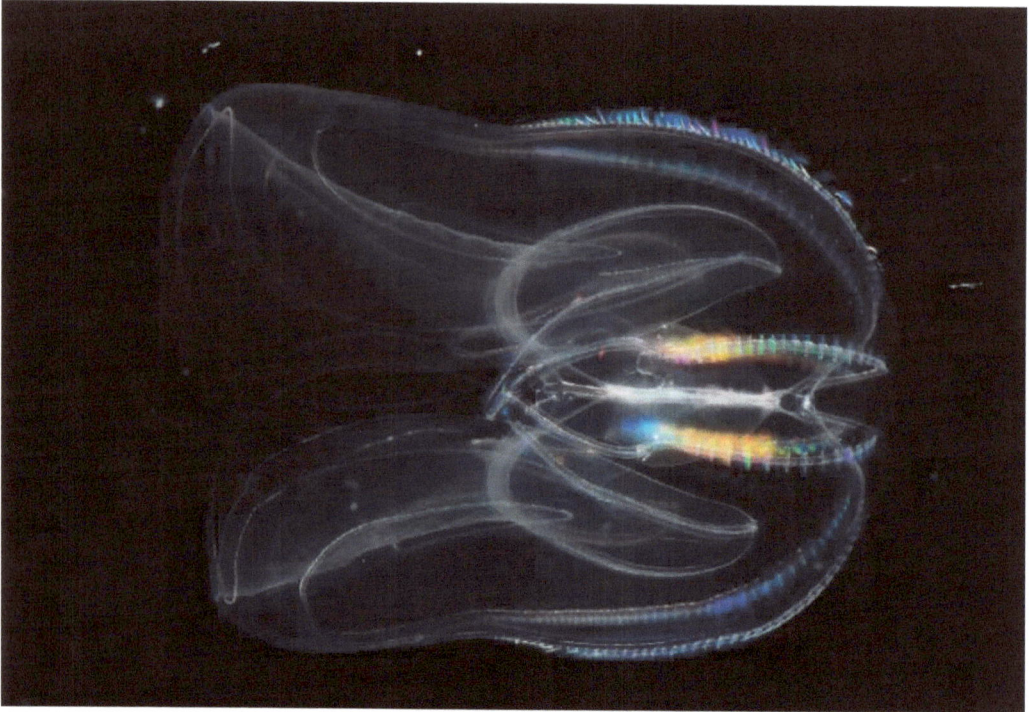

Fig. (17). *Mnemiopsis leidyi.*

Bioluminescence: It produces an intracellular bioluminescence that appears as flickering thin blue lines along its comb rows [38]. Interestingly its bioluminescence begins even in its embryonic stage. Further, its ciliated combs are iridescent by day and they may glow green by night.

The bioluminescence intensity of this species showed considerable seasonal fluctuations. Low bioluminescence values were observed in the spring period. Further, shortest flashes were registered in February–March, making 0.79–1.32 s and more prolonged luminescence duration is observed in August–September and it achieves 2.77–3.46 s. Furthermore, its light-emission characteristics rose with a peak in August [87].

Ocyropsis fusca

Image credit: Wikipedia

Ocyropsis fusca (Fig. **18**) is a tropical, luminescent species. No other information is available.

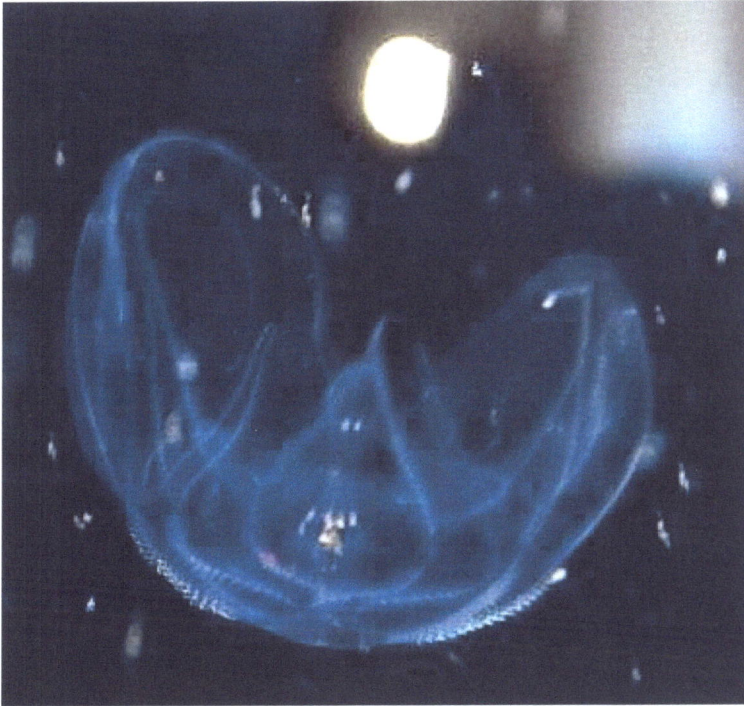

Fig. (18). *Ocyropsis fusca.*

Ocyropsis maculata

Image not available

It is found in the outer continental shelf and slope of Western Central Atlantic areas such as the USA and Mexico. In this luminescent species, four distinct brown spots are seen on its large, prominent oral lobes. The body is compressed. The color of lobes is whitish, translucent. It attains a maximum length of 10 cm.

Thalassocalyce inconstans

Image not available

It is an inshore and offshore species occurring at depths of 0 - 785 m in the Eastern Pacific, Western Central Atlantic and the Mediterranean. This luminescent species has an extremely fragile bodythat may reach 15 cm in width. The body of this species, however, is much shortened in its oral-aboral view. It has short comb-rows on the surface furthest from the mouth and they are originating from near the aboral pole. It has no tentacles hanging outside the bell. It may have a hemispherical d shape when fully expanded but it possesses a bi-radial shape when it is partially contracted into a "two-globe form".

Tizardia phosphorea

Image not available

This tropical, luminescent species is found distributed in Western Central Pacific.

Bioluminescence: Though luminous organs have been recognized in this species, direct observation of its luminescence is lacking [13].

Velamen parallelum

Image not available

It is a cosmopolitan species found in the outer continental shelf and slope. It largely resembles Cestum sp. It lacks sub-tentacular comb rows. Its sub-tentacular meridional canals arise from adradial canals at the edge of wings. The body is transparent and reaches a maximum length of 15 cm.

Bioluminescence: its luminescence is largely due to its green- fluorescent proteins [57].

Class: Nuda

Beroe abyssicola

Image credit: Wikimedia

Beroe abyssicola (Fig. **19**) is a deepwater species found distributed from Subtropical to polar in the Eastern Pacific, Black Sea, Northeast Atlantic and the Arctic regions such as Canada, Russia and USA. It has a very different body plan from other Ctenophores, namely the absence of tentacles in its life cycle. It has a muscular, flat, and cylindrical body that can grow up to 7 cm long. The body is more opaque its coloration is red or purple.

Bioluminescence: Like other ctenophores, it has a rainbow effect on its comb rows and this is due to light refraction. However, it also possesses bioluminescence which is caused by calcium-activated photoproteins, similar to hydromedusae. This photoprotein is called berovin which differs from that of hydromedusa in that it is sensitive to visible and UV light. Like other ctenophores, the bioluminescent displays of this species were all multiple flashes that propagated along the comb rows [60].

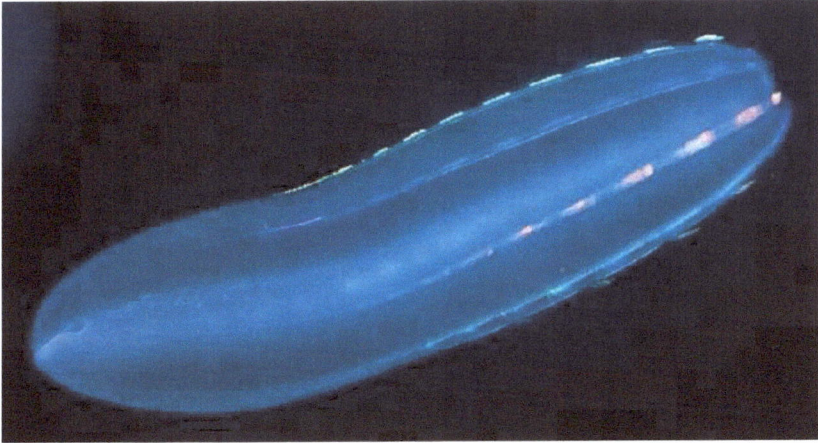

Fig. (19). *Beroe abyssicola.*

Beroe cucumis

Image credit: Mark Norman/Museum Victoria, Wikimedia

Beroe cucumis (Fig. **20**) is commonly called as "pink slipper comb jelly" and has cosmopolitan distribution *i.e.* tropical to polar. Its depth range is 0 - 880 m. It has a transparent, sac-like body, which is somewhat compressed reaching a maximum length of about 15 cm. It has a wide mouth at one end. The body has eight longitudinal rows of cilia that extend from the aboral end (*i.e.* opposite end to the mouth). Cilia are arranged on short transverse plates. The general body colour is pink, especially along the rows of cilia. A figure of eight shaped rings of small papillae is seen around the aboral tip.

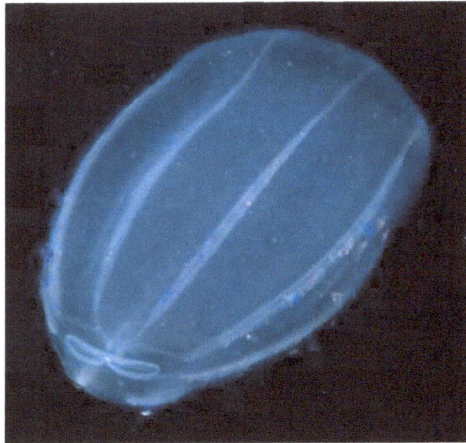

Fig. (20). *Beroe cucumis.*

Bioluminescence: The eight rows of locomotory cilia which run along the body of the animal are more easily visible than the rest of the body surface, due to the stronger light scattering taking place on these protrusions. Further, the "comb"-rows of this species appear to be brightly colored, showing an iridescence. This is however, not related to any bioluminescence but it is a selective reflection from a two-dimensional photonic crystal [94].

Beroe forskalii

Image credit: Flickr

Beroe forskalii (Fig. **21**) is found distributed at depths of 0 - 630 m in Tropical to polar regions of the Atlantic, the Mediterranean, Pacific and Antarctic Atlantic. The body of this species is shaped like a compressed cone. There are eight rows of combs that run 3/4 the length of the body. The mouth opening is wide, with large, half circle-shaped lips. It is slightly pinkish in adults. Its height is up to 20 cm.

Fig. (21). *Beroe forskalii.*

Bioluminescence: It produces a bright luminescent display when disturbed. But its rainbow colors are not bioluminescence and are merely diffraction acting on the ambient light [95]. However, it is also reported that this species can express bioluminescence internally, in cascading waves [96].

Beroe gracilis

Image credit: Christian Coudre (Reproduced with permission)

Beroe gracilis (Fig. **22**) is a shallow water species distributed from subtropical to temperate regions of the Eastern Pacific and the Atlantic Ocean at a depth range of 12 - 200 m. The body of this species is slender cylindrical with slight lateral compression. Ciliary comb-rows extend from the aboral pole to about three-quarters of the distance towards the mouth. A row of branched papillae in the form of a figure-8 is seen around the pole plate at the aboral pole. Its four meridional canals have no side branches. Adult has a milky appearance and some specimens may be pinkish. The size of an adult is up to 30 mm high. Its luminescence largely resembles that of *Beroe forskalii.*

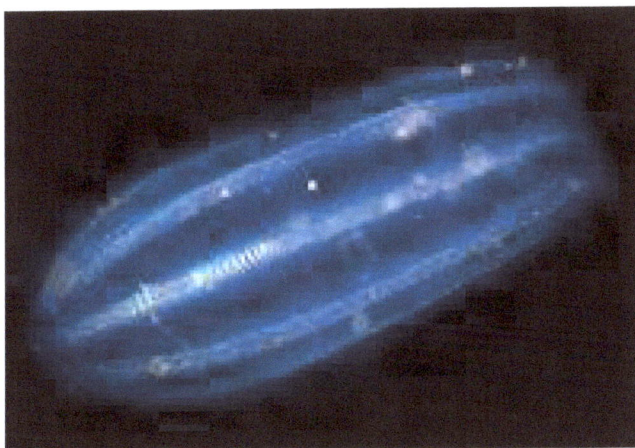

Fig. (22). *Beroe gracilis.*

Beroe ovata

Image credit: CristianChirita, Wikimedia

Beroe ovata (Fig. **23**) is commonly called as "brown comb jelly" and this subtropical species occurs at a depth range of 0 - 1719 m in Indo-West Pacific, Atlantic Ocean, and the Mediterranean. The body of this species grows to a total length of 16 cm. It is fairly oval or cylindrical. Its body wall is composed of a gelatinous mesoglea . It is pale blue or s pale pink. On its exterior surface, there are eight longitudinal rows of cilia that form the "combs". This species has no tentacles and its internal gastric cavity is connected to a network of canals so as to form a meshwork in the mesoglea.

Bioluminescence: In this species, its photoprotein and luciferin quantity increasehave been observed in adult individuals. Owing to this, its bioluminescence quantum output is minimum at the early stages of development, and maximum at late [97]. This species showed the highest index of bioluminescence in the summer period and minimal index in the winter-spring period. Further, in this species, the bioluminescence reaction optimum was achieved under the temperature of 22 °C while its minimum was registered under the temperature of 10 °C [87].

Fig. (23). *Beroe ovata.*

Nepheloctena sp.

Image credit: Griswold, National Oceanic and Atmospheric Administration.

Nepheloctena sp. (Fig. **24**) is an undescribed pelagic ctenophore (Anon., https://thereaderwiki.com/en/Ctenophora) and this luminescent species is found in deeper waters. It has Tortugas red coloration and is with trailing tentacles and distinct side branches, or tentilla.

Fig. (24). *Nepheloctena sp.*

Bioluminescence: it has been reported to produce bright comb row flashes [68].

CONCLUSION

At present, quite great number of works have been devoted to the physiology and ecology of different ctenophore species, but bioluminescence in general, and the studies of the light-emission parameters in particular still remains to be not much studied in more than 100 species living in a wide range of temperatures . Further, the influence of different environmental factors on the bioluminescence parameters and connection between the organism's physiological indices and its luminescence need to be studied. In connection with the above mentioned, it is considered to be extremely important to continue the investigation of the light-emission in the ctenophores of the world seas in order to make use of such luminescent ctenophores for biomedical and biotechnological applications.

CHAPTER 7

Bioluminescent Planktonic Marine Annelids

Abstract: This chapter deals with the luminescent species of planktonic annelids such as *Tomopteris* spp and syllid larva, the emission maxima in observed species of *Tomopteris*, the description of luminescent species of planktonic annelids and their mechanism of bioluminescence.

Keywords: *Chaetopterus*, Emission maxima, *Tomopteris*.

INTRODUCTION

The annelids (also called "ringed worms") (Phylum: Annelida) are segmented worms, with about 17,000 species, including polychaetes, clitellates, ragworms, earthworms, and leeches. They are found in marine environments from tidal zones to hydrothermal vents, in freshwater, and in moist terrestrial environments. Among the 13 known families of annelids, eight of which are marine, with both benthic and pelagic animals producing bioluminescence. These eight marine families include the well-studied, charismatic, displays from Chaetopteridae, Tomopteridae, and Syllidae. Some notable examples are the benthic "fireworms"(*Hermodice* spp) which use luminescence as part of their mating display, and planktonic *Tomopteris*, which makes yellow light. On land, some oligochaetes (earthworms) produce luminous secretions. Among the marine polychaetes, the species of *Alciopina* and *Rhynchonereella* (Alciopidae), *Tomopteris* (Tomopteridae), *Chaetopterus* (Chaetopteridae), *Poeobius* (Flabelligeridae) and larvae of *Odontosyllis* (Syllidae) are the planktonic, luminescent marine organisms. However, the luciferin and luciferase of the benthic adult fireworm *Odontosyllis* have only been identified and purified though the light emission process within annelids and seems to be distinct in each family. Among the wide range of taxa, annelid bioluminescence in general and planktonic annelid bioluminescence is one of the least studied. This chapter deals with the findings relating to planktonic, luminescent annelids.

Of the holopelagic polychaete families of the phylum Annelida, only 13 species have been reported to be luminescent (Table **1**). Of these species, the members of

Tomopteridae are well-known. Most of these tomopterids are known to produce a golden yellow bioluminescent light under nervous control,b species that emit blue light have also been reported in this family [98]. Though photogenic organs have not been reported in the Alciopidae, some species of this family have been reported to be luminescent. In general, the structure and function of the photogenic organs and the associated luminescent products are incompletely known in these pelagic polychaetes [99].

Table 1. Bioluminescent planktonic annelids [13]

Class	Order	Family	Species
Polychaeta	Phyllodocida	Alciopidae	*Alciopina* sp., *Rhynchonereella* sp.
-	-	Syllidae	*Odontosyllis* sp. (larva)
-	-	Tomopteridae	*Tomopteris pacifica, T.carpenteri, T.helgolandica, T.nisseni, T. septentrionalis, T.planktonis, T.mariana, T.nationalis*
-	Spionida	Chaetopteridae	*Chaetopterus pugaporcinus*
-	Terebellida	Flabelligeridae	*Poeobius meseres*

Bioluminescence in tomopterids: The holoplanktonic tomopterids (Annelida: Polychaeta) have a maximum size of 10 cm and are found distributed worldwide from the near-shore waters to about 3000 meters in depth. A total of twelve species of *Tomopterisviz.*: *Tomopteris anadyomene, Tomopteris apsteini, Tomopteris elegans, Tomopteris helgolandica, Tomopteris kefersteini, Tomopteris krampi, Tomopteris mariana, Tomopteris nationalis, Tomopteris nisseni, Tomopteris planktonis, Tomopteris rolasi,* and *Tomopteris septentrionalis* have been reported topossess photogenic organs in their parapodial glands and emit golden-yellow colored light (Fig. **1**) with an emission maximum (λmax) of 565-570 nm. In the yellow-colored light emitting system of these animals, a photoprotein probably activated by superoxide anion is believed to be involved. Further, the presence of cœlenterazine (28 pmol in one specimen) and a fluorescent emitter has also been documented from the tissue homogenates of these tomopterids. The bioluminescent system components of tomopterids have also been reported to be Aloe-emodin/luciferase . Although the luminescence of these tomopterids was thought to be intracellular, a secretory activity has also been suggested in several species of *Tomopteris* which release the luminous exudate while trying to escape from a collecting device [21]. It is reported that the luminescence emission of polychaetes is due to the oxidation reaction of luciferin and oxygen catalyzed by the enzyme luciferase or species-specific protein [100].

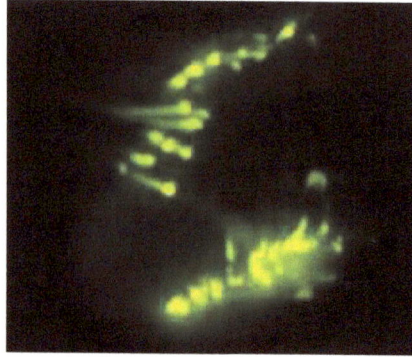

Fig. (1). Yellow light produced by *Tomopteris* spp.

Image credit: The Oceanography Society (Reproduced with permission)

Emission maxima in luminescent marine planktonic annelids: The emission maxima (λmax) of the different species of luminescent marine planktonic annelids have been reported to vary from 450 to 574 nm (Table **2**).

Table 2. Emission maxima (λmax) of luminescent planktonic marine annelids.

Species	λmax	Refs.
Tomopteris carpenteri	565-570nm	[21]
-	574nm	[98]
Tomopteris helgolandica	565-570 nm	[21]
-	573nm	[98]
Tomopteris nationalis	550, 570nm	[102]
Tomopteris nisseni	565nm	[98]
Tomopteris pacifica	565-570 nm	[21]
-	549nm	[98]
Tomopteris planktonis	450 nm	[21]
Tomopteris septentrionalis	565-570 nm	[21]
-	557nm	[98]
Tomopteris spp.	565nm	[98]
-	493nm	[98]
Poeobius meseres	495nm	[101]

LUMINESCENT SPECIES OF PLANKTONIC ANNELIDS

Family: Alciopidae

Alciopina sp.

Image not available

The holoplanktonic alciopids are cosmopolitan in distribution and are epipelagic, inhabiting mainly the upper 50 meters of the water column. These are long,slender wormswith a short body, large eyes that are laterally directed, rounded prostomium,. one ventral pair of palps, and three antennae. They also have six pairs of tentacular cirri, and simple capillary Chaetae.

Bioluminescence: Though light-producing photogenic organs are not present in these alciopids, some species have been reported to be luminescent [99].

Rhynchonereella sp.

Image not available

It is cosmopolitan in distribution, and can occur in the upper and midwater columns. The body of this species is elongated and slender. The prostomium is produced anteriorly to eyes, with three antennae (one median and two dorsal), and two ventral palps. The first three segments have 4–5 pairs of tentacular cirri and the anterior parapodia are well developed. Lastly, there are three types of Chaetae.

Bioluminescence: Though light-producing photogenic organs are not present in these alciopids, some species have been reported to be luminescent [99].

Family: Syllidae

Odontosyllis sp. (larva)

Image not available

These luminescent *Odontosyllis* larvae may inhabit algal mats, sediments, or seawater. These larvae possess elongated bodies with numerous segments and their sizes range from 5 to 10 mm.

Bioluminescence: These larvae have been reported to emit a strong flash of yellow-green colored light,believed to be due to the self-internal luminescence pathway [100].

The larvae of *Odontosyllis* spp. emit an intense glow of luminescence on the addition of freshwater or with gentle disturbance. Though flashes of internal luminescence were observed for some time, the addition of freshwater yielded a continuous glow that lasted for a few minutes [103].

The tomopterids are dorsoventrally flattened holoplanktonic polychaetes and are mainly characterized by their modified parapodia, conspicuous palps, and long parapodial cirri on the second segment. Most species are transparent or have little pigmentation. While most of the species are small, some species may attain a large size of 10 cm or more.

Tomopteris anadyomene

Image not available

The body of this luminescent species consists of a head, a cylindrical segmented body, and a tail. The head has a prostomium (part in front of the mouth opening) and a peristomium (part around the mouth), and it bears paired appendages *viz.* palps, antennae, and cirri.

Tomopteris (Johnstonella) apsteini

Image not available

It is found distributed in the northern Pacific and is distinguished by the absence of chromophile glands in the third pair of parapodia and rosettes.

Tomopteris carpenteri

Image credit: Hauke Flores, AWI (Reproduced with permission)

Body of this luminescent species (Fig. **2**) is very large and considered to be largest species of this genus . It is found distributed throughout the water column, including the deep waters in Antarctica's Southern ocean. It is adorned with its characteristic alternating red and transparent bands. Pinnules are oval, frilly and they extend on to the distal ends of the parapodial trunks.

Bioluminescence: It can shoot bioluminescent sparks off their feet when threatened.

Fig. (2). *Tomopteris carpenteri .*

Tomopteris(Johnstonella) helgolandica

Image credit: Fisheries and Oceans Canada, WoRMS

This luminescent species (Fig. **3**) is commonly found near the coast in the North Atlantic Ocean and in the Mediterranean. This worm has a short and broad body which is striated with dark brown lines. It has a maximum of 34 segments, in addition to a tail. Its parapodia which are without chaetae are elongated and are with rosette organs in the tips of both branches. Prostomium is with two short, divergent antennae; and two eyes with lenses and two ciliated nuchal organs. Tail may be one third of the total body length.. In life, it is yellow . Body length can attain a maximum length of 4 cm, including tail.

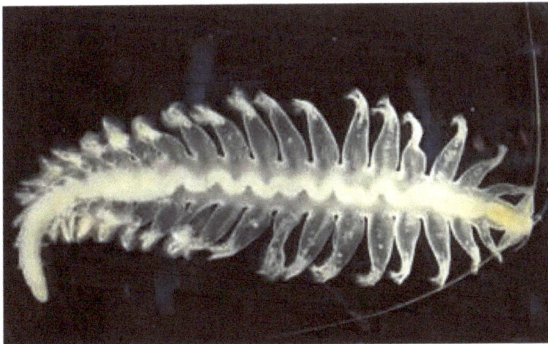

Fig. (3). *Tomopteris(Johnstonella) helgolandica .*

Bioluminescence: Its yellow-pigmented parapodial glands possess bioluminescent properties. It produces yellow light flashes *via* activation of nicotinic cholinergic receptors and a calcium-dependent intracellular mechanism involving L-type calcium channel [104].

Tomopteris krampi

Image not available

This luminescent species is found distributed in southern Atlantic and NE pacific. Its pinnules ae found extended on the distal ends of parapodia trunks. Hyaline glands are seen in the apices on dorsal and ventral Pinnules from the third parapodia. Individuals of this species can grow to a maximum of 15 mm.

Tomopteris mariana

Image not available

Bioluminescence in tomopterids was first reported in the parapodial pigmented rosettcs (glandular organs) of this species [98]. It is suggested that the yellow cells of the grandular organs may be photocytes that elaborate and accumulate lipoid products which would be the support of the photogenic substances. These transparent cells may be either a lens structure or nutrition cells connected with the organ. In this species, a nerve ganglion is to be found at the base of the transparent cells with fibres running between them to the putative photocytes [104].

Tomopteris nationalis

Image not available

This species is found distributed in Northern Atlantic, Western Indian Ocean and the Mediterranean Sea It has also been recorded from Europe and the Arabian Sea. It is a small species with a long tail. Its stout anterior cirri may be ost as long as its prostomial horns. Rosettes which are seen near the tips of the rami in the mid-body region are also present on the trunks of the first two pairs of parapodia.

Bioluminescence: This species has been reported to possess two fluorophores one appearing yellow-green with ultraviolet excitation, the other yellow-orange. The yellow-orange fluorescent material which is located near the photocytes (light-emitting cells) are believed to be involved in the bioluminescence. The compound present in this fluorescent yellow-orange material of all tomopterids has been identified as aloe-emodin (Fig. **4**), a polyhydroxyl-substituted anthraquinone [102]. An increase of the bioluminescent potential has been observed in this species during its sexual activity period [21].

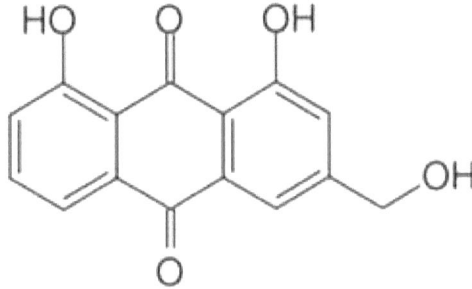

Fig. (4). Aloe-emodin.

Tomopteris nisseni

Image credit: Scripps Institution of Oceanography

This luminescent species (Fig. **5**) is a deep-sea planktonic worm found distributed in Western Central Atlantic and the Mediterranean Sea areas of Central America to northern Brazil. The body of this species is large and it has a long tail. Pinnules are reduced to a fringe bordering the parapodia. Hyaline glands are in a variable position. Its cirriform appendages of the second segment may exceed the total length of the body. Gonads are present in the dorsal rami only, the most anterior is in the first or second parapodial segment.

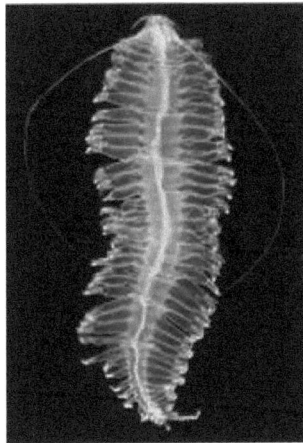

Fig. (5). *Tomopteris nisseni* .

Bioluminescence: This species is one of the few creatures on the planet to produce yellow bioluminescence [105]. It has been reported to produce bright light flashes *via* activation of nicotinic cholinergic receptors and a calcium-dependent intracellular mechanism involving L-type calcium channels.

Tomopteris   (Johnstonella)   pacifica   (= *Tomopteris   elegans;   Tomopteris kefersteini*)

Image credit: Fisheries and Oceans Canada, Moira Galbraith, WoRMS

This species (Fig. 6) is commonly called as "tailed Pacific transparent-worm" and is found distributed between e surface and a depth of 75 m in Indo-Pacific and the Mediterraneanareas such as Mexico, Peru and Galapagos Islands. It is a small, delicate worm with slender parapodia and well-developed anterior cirri. Its hyaline glands are restricted to the dorsal pinnules of its third and fourth pairs of parapodia. Gonads are present in the dorsal rami only. It has no tail . It attains a maximum length of 4 cm.

Fig. (6). *Tomopteris (Johnstonella) pacifica* .

Bioluminescence: Its light spots have been reported to vary much in size, but many of them are smaller than 20 µm in diameter, which is smaller than the overall size of the rosette-like glands. It is suggested that the source of its light emission may be from its grains or from the secretion of its yellow-pigmented bodies [104].

Tomopteris planctonis

Image not available

It is a blue emitter and the light seems to be produced from its yellow-pigmented parapodial glands. Its blue bioluminescence could act as a deterrent signal [104].

Tomopteris rolasi

Image not available

This luminescent species is found distributed in Indo-West Pacific and Eastern Central Atlantic regions such as Sao Tome and China. In this species. Its pinnules are oval. Rosette glands are present in the trunks of the first two parapodia and near the tips of both rami. Its chromophil glands are rounded and are situated ventrally from the third parapodia.

Bioluminescence: As in *Tomopteris mariana*.

Tomopteris septentrionalis

Image credit: Fisheries and Oceans Canada, Moira Galbraith, WoRMS

This pelagic but mainly oceanic species (Fig. **7**) is found distributed worldwide and is common in the North Sea to Baltic, Faeroes and Iceland. Body of this species is short with long 17-24 long parapodia. Its transparent body is posteriorly tapering but without tail. There are no rosette organs. Prostomium has 4 small incisions anteriorly. Eyes are large. Anterior cirri are absent. Gonads are present in the dorsal rami only. It may attain a maximum length of 30 mm or more.

Fig. (7). *Tomopteris septentrionalis .*

Bioluminescence: The light spots of this species varied much in size, but many of them were found to be smaller than 20 µm in diameter, which is smaller than the overall size of the rosette-like glands. It is suggested that the source of its light emission may be from its grains or from the secretion of its yellow-pigmented bodies [104].

Chaetopterid Worm

Chaetopterus pugaporcinus

Image credit: Kim Fulton-Bennett

This is the only planktonic species of *Chaetopterus* (Fig. **8**) found in the deep (875- 1200 m) mesopelagic waters off (midwaters) Monterey Bay, California. All specimens of this species collected exhibited a combination of both adult and larval characteristics. This worm is round in shape and is about 10 - 20 mm in length . This species has a strong resemblance to a disembodied pair of buttocks. Because of this, its Latin name which roughly translates to "resembling a pig's rear. These worms are neutrally buoyant, and are found floating along with their mouthparts facing downward, and their hind parts towards the ocean surface. The worm has a segmented body, but as the middle segments are highly inflated, it is giving the animal a round shape.

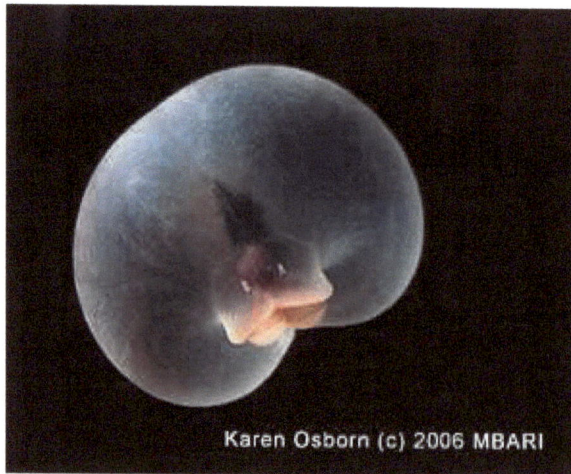

Karen Osborn (c) 2006 MBARI

Fig. (8). *Chaetopterus pugaporcinus* .

Bioluminescence: This species has been reported to produce light in two forms *viz.* bright blue light and glow associated with its bioluminescent green particles . The blue light is from its peristomium/prostomium after direct physical stimulation. This area glowed for 3–6 s, then abruptly extinguished. The other light source *viz.* the minute green, bioluminescent particles were found spewed from the middorsal ciliated groove or surrounding area, and the posterior end. These small glowing specks were dispersed throughout the mucous cloud produced light at the same time, and it glowed clearly for 1–2 s before it faded slowly. The mucus and bioluminescent particles were believed to be forced away from the body by as much as two body lengths [98]. The blue light-emitting of this species is due to the involvement of ferritin and riboflavin in this Chaetopterus photoprotein luminescence reaction. Further in this reaction, iron was an important cofactor [106].

Flabelligerid Worm

Poeobius meseres

Image credit: MBARI ROV Tiburon and Karen Osborn (Reproduced with permission)

It is actually a holopelagic species (Fig. **9**). As it is gelatinous, it is included as a member of plankton (Poupin, 1999). It is the sole recognized luminescent representative of Poeobiidae (Annelida). It has been recorded from the eastern Pacific at depths of 350 –1300 m.

Fig. (9). *Poeobius meseres .*

The worm is up to 27 mm in length and it has 11 poorly defined segments, Body is largely gelatinous with a thick mucus sheath, and segmentation is therefore not clearly visible. Its anterior end bears retractable pale green "tentacles," which consist of a pair of grooved palps and branchiae.

Bioluminescence: It has been reported to generate light along the whole length of the body [101]. Its detailed spectral properties are not however known.

CONCLUSION

Currently, bioluminescence is primarily known from members of the subfamily Polynoinae which includes the benthic scale worms. Further, even though a number of bioluminescent polynoids are known, the chemical mechanism of light emission in these organisms is not clearly understood. Furthermore, intensive studies are needed to estimate the diversity of planktonic, luminescent polychaetes in the world seas. The elucidation of the biochemical pathway involved in bioluminescence can contribute to understanding the ecology of planktonic, luminescent polychaetes and help toward the development of biotechnological tools. As many photoproteins found in bioluminescent organisms are currently used as luminescent probes in clinical assays, the role of luminescent polychaetes may also likely to be significant.

CHAPTER 8

Bioluminescent Chaetognaths

Abstract: This chapter deals with the luminescent species of Chaetognatha *viz.* *Caecosagitta macrocephala* and *Eukrohnia fowleri*,their emission maxima and mechanism of bioluminescence.

Keywords: *Caecosagitta macrocephala*, Emission maxima, *Eukrohnia fowleri*.

INTRODUCTION

The phylum Chaetognatha, commonly known as arrow worms, contains about 200 species and about 80% of them are planktonic, bilaterally symmetrical, coelomate and worm-like organisms. Chaetognaths may be found in all marine environments, from tropical surface waters and shallow tide pools to the deep sea and polar regions. While most chaetognaths are transparent and torpedo-shaped, some deep-sea species are orange. Among the arrow worms, very few species like *Caecosagitta macrocephala* and *Eukrohnia fowleri* have been reported to be luminescent by using the luciferin coelenterazine to make their light. This chapter deals with such luminescent species of chaetognaths and their light emission associated with luciferin–luciferase reaction.

The phylum Chaetognatha (meaning bristle-jaws) are commonly known as arrow worms. These marine planktonic organisms are distributed worldwide and are found in all marine environments from surface tropical waters to the deep sea and polar regions. Most chaetognaths are transparent and are torpedo-shaped, but some deep-sea species are orange. They range in size from 2 to 120 millimetres. Though there are more than 120 modern species of Chaetognatha, only two species *viz.* *Caecosagitta macrocephala* and *Eukrohnia fowleri* are known to luminescent. In these species, the bioluminescent organs are present on their fins. These chaetognatha use luciferase and coelenterazine to shed light during an escaping response [107].

Caecosagitta macrocephala

Image credit: WoRMS

This deep-sea (200 - 5000 m) luminescent species (Fig. **1**) has a cosmopolitan distribution and is commonly seen in the north-western Pacific and the centre-east of southern Atlantic Oceans. It has been reported to have a very wide distribution that ranges from the Subantarctic to the Subarctic Ocean. It has a large head. Along with its eyes, its gut or intestine has orange pigmentation and a luminous organ that flashes light due to bioluminescence, unlike other species of Sagittidae. Its body length may range from 16 to 22 mm. Its hooks and teeth are deep browns. It is worth mentioning that the individuals living near the bottom may be morphologically slightly different from those in the upper water column. Animals of the latter may possess foamy collarette on the entire body, long anterior fins and no traces of eyes.

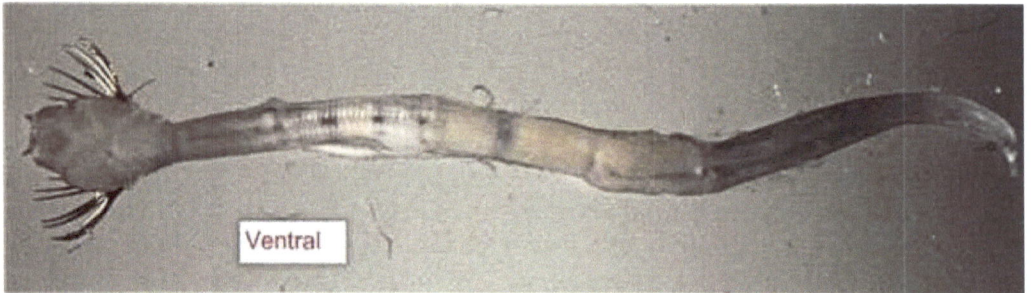

Fig. (1). *Caecosagitta macrocephala.*

Eukrohnia fowleri

Image not available

It is a bathy or mesopelagic oceanic species found mainly below 700 meters throughout the world. The body of this species is firm and broad with a large head . It has partially rayed long fins that originate at the level of the ventral ganglion. The alimentary diverticula are absent,the gut is usually reddish, and the eyes are provided with triangular pigment spots. The animals are brown in coloration with straight tips. There are 2-31 teeth, with short and broad ovaries, large and ovoid seminal vesicles, and a maximum body length of 4 cm.

Bioluminescence: In these species, their bioluminescence is based on a coelenterazine-luciferase reaction [13]. The luminescent organ of *Caecosagitta macrocephala* (Fig. **2**) is located on the ventral edge of each anterior lateral fin, whereas that of *Eukrohnia fowleri* (Fig. **3**) runs across the center of the tail fin on both dorsal and ventral sides [108]. Further, they reported that the bioluminescent organs of both species consisted of hexagonal chambers containing elongated ovoid particles called the organelles, which are believed to hold bioluminescent materials. The transmission electron microscopy of these particles from *Caecosagitta macrocephala* revealed a densely packed paracrystalline matrix punctuated by globular inclusions, which seemed to correspond to luciferin and luciferase, respectively. Further, both species used unique luciferases in conjunction with coelenterazine for light emission. While the luciferase of *Caecosagitta macrocephala* became inactive after 30 minutes, luciferase of E. fowleri was found to be highly stable. Although *Caecosagitta macrocephala* had about 90 times fewer particles than *Eukrohnia fowleri*, it had a more or less similar bioluminescent capacity (total particle volume) due to its larger particle size. *In situ* observations of *Caecosagitta macrocephala* revealed that the released luminous particles formed a cloud.

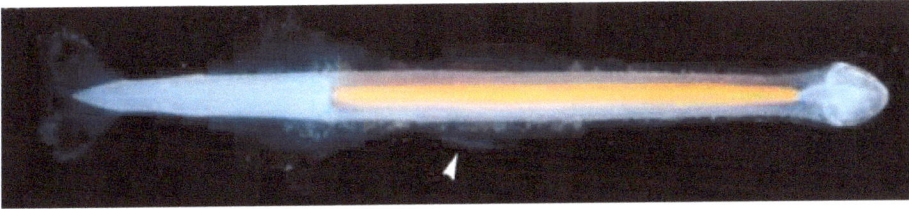

Fig. (2). Bioluminescent organ of *Caecosagitta macrocephala* (arrow-locating).

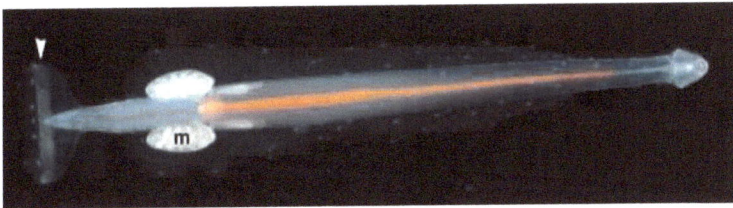

Fig. (3). Bioluminescent organ of *Eukrohnia fowleri* (arrow-locating). Image credit: Erik V. Thuesen, Ph.D. (Reproduced with permission)

CONCLUSION

The functional mechanics of light emission in chaetognaths remain largely unknown. Although the photocytes of luminous organs are stimulated to luminescence by nervous conduction in other organisms, the luminescence mechanism in these chaetognaths is unique in the fact that these photocytes will flash and glow even once they detached from the chaetognath. It indicates that these photo particles are passively disrupted when exposed to seawater. Further, intensive research on the diversity of luminescent chaetognaths and the characterization of the novel luciferases from these chaetognath species will thereforebe a fascinating glimpse into the evolution of coelenterazine-based bioluminescence.

CHAPTER 9

Bioluminescent Marine Crustaceans

Abstract: This chapter deals with the luminescent species of planktonic marine crustaceans such as ostracods, copepods, amphipods, mysids, euphausiids and decapods; and their emission maxima and mechanism of bioluminescence.

Keywords: Amphipods, Copepods, Emission maxima, Euphausiids, *Lucifer*, Mysids, Ostracods.

INTRODUCTION

Marine crustaceans form a very diverse group of invertebrate animals including planktonic members such as krill, copepods, amphipods, nektonic lobsters,shrimp, benthic crabs and more sessile creatures like barnacles. These crustaceans play a key role as an important source of nutrition for a wide range of marine vertebrates such as fish, birds and mammals (seals, whales). Most planktonic crustacean groups (with the exception of isopods) such as copepods, ostracods, amphipods, decapod shrimp and euphausiids (krill) possess luminous members. Uniquely, three of the major marine luciferins *viz.* ostracod-type luciferin, dinoflagellate-type luciferin, and coelenterazine are used in various crustaceans in the chemical reactions associated with the production of light. Further, crustaceans have been reported to produce coelenterazine and they are believed to be the major source for coelenterazine in the sea. Among crustaceans, luminous species are especially remarkable in the copepods, shrimps, and ostracods. Some shrimp (*Hoplophorus*) emit a luminous secretion from luminous organs, while others possess true light organs (photophores), which may be internal or superficial and glow steadily. These photophores consist of a lens, reflector, and light-emitting photogenic cells.

BIOLUMINESCENT MARINE OSTRACODS

Ostracods occur in all oceans from polar to tropical waters and at all depths (surface to about 4,000 meters). About 50% of the hither-to described 300 species of marine ostracods as luminescent (Table **1**). Like copepods, these organisms possess luminescent glands which release light-producing substances into the water. Among the ostracods, the species of *Cypridina* possess a large luminous gland in which two types of gland cells are present. While one type secretes the substrate "luciferin" the other secretes the putative photoprotein, luciferase. On stimulation, these organisms squirt both the luciferin and luciferase into seawater. The mixing of these two substances produces a blue cloud of luminescence in the seawater [13].

Table 1. Bioluminescent marine ostracods [13].

Class	Order	Family	Species
Ostracoda	Myodocopida	Cypridinidae	*Cypridina americana, C. dentata, C. noctiluca, C.serrata, Vargula antarctica, V. bullae, V.harveyi, V. hilgendorfii, V. norvegica, V. tsujii*
-	-	Halocyprididae	*Conchoecia acuminata, C. alata, C. ametra, C.atlantica, C. belgicae, C.bispinosa, C. borealis, C. concentrica, C. curta, C . daphnoides, C. echinate, C.elegans, C. haddoni, C.hyalophyllum, C. imbricata, C. kampta, C. lophura, C. loricata, C. macrocheira, C.magna, C. oblonga, C. parthenoda, C.procera, C. rhynchena, C.secernenda, C.spinifera, C. spinirostris, C.subarcuata, Euconchoecia chierchiae*

Emission Maxima in Luminescent Marine Planktonic Ostracods: The emission maxima (λmax) of the different species of luminescent marine planktonic ostracods have been reported to vary from 459 to 481nm (Table **2**).

Table 2. Emission maxima (λmax) of luminous marine ostracods.

Species	λ max	Ref
Conchoecia imbricata	474nm	[13]
Conchoecia secernenda	481nm	[13]
Vargula antarctica	475nm	[109]
Vargula hilgendorfii	459-465nm	[10]
-	465-469nm	[13]

Bioluminescent Marino Ostracods

Family: Cypridinidae

Cypridina noctiluca (= *Pyrocypris noctiluca*)

Image credit: Dr. Yuichi Oba, Ph.D. (Reproduced with permission)

This luminescent species (Fig. **1**) is widely distributed in the coastal waters of the western Pacific (from southern Japan to Hawaii, Australia, and Southeast Asia) and in the Indian Ocean. It is a large ostracod, measuring about 2 mmand has a cylindrical tail, posteriorly and anteriorly with a very shallow incisure. The dorsal and ventral margins are not broadly rounded, the surface of the shell is smooth, and the uropodal lamellae haveeight claws.

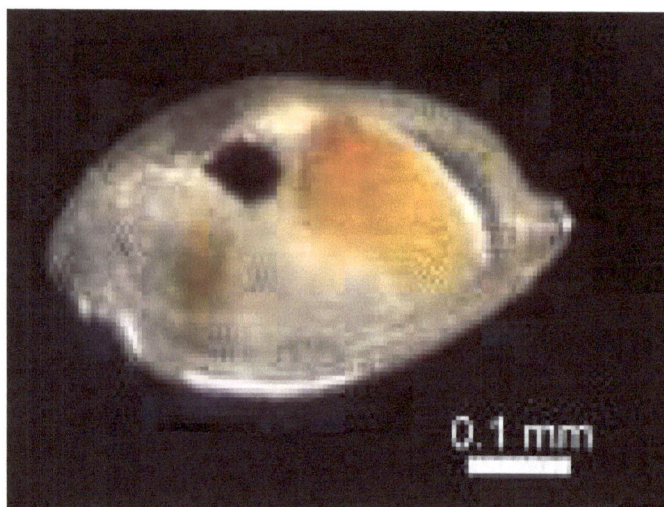

Fig. (1). *Cypridina noctiluca.*

Bioluminescence: The luminous ostracods of the genus *Cypridina* (family Cypridinidae), commonly called sea fireflies, produce blue light with an emission maximum (λmax) ranging between 448 and 463 nm. In the bioluminescent reaction of *Cypridina* spp. the Cypridina luciferin (now cypridinid luciferin) (also known as Vargulin) is oxidized in the presence of Cypridina luciferase (CLase) (now cypridinid luciferase) and molecular oxygen (oxidation step), followed by generation of the oxyluciferin in the excited state (excitation step) and subsequent change to the ground state with light emission (light production step). The luminescent reaction in cyprinids is detailed below [110].

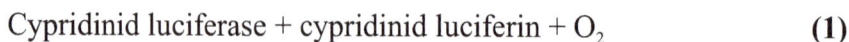

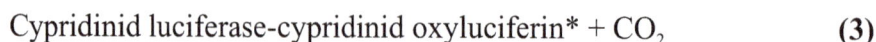

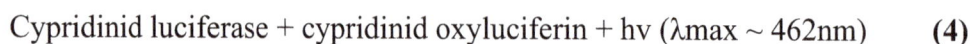

$$\text{Cypridinid luciferase + cypridinid luciferin} + O_2 \tag{1}$$

$$\text{Cypridinid luciferase-cypridinid luciferin-dioxetanone intermediate} \tag{2}$$

$$\text{Cypridinid luciferase-cypridinid oxyluciferin* } + CO_2 \tag{3}$$

$$\text{Cypridinid luciferase + cypridinid oxyluciferin + hv } (\lambda\text{max} \sim 462\text{nm}) \tag{4}$$

Cypridina noctiluca has been reported to synthesize Cypridina luciferin de novo (anew) from three amino acids *viz.* L-tryptophan, L-isoleucine, and L-arginine [4]. The luciferase of this species has been utilized for biochemical and molecular biological applications, including bioluminescent enzyme immunoassays, far-red luminescence imaging, and high-throughput reporter assays [111].

Cypridina serrata (= Pyrocypris serrata)

Image not available

It is a subtropical species found distributed in the Western Pacific region, particularly in China.

Bioluminescence: This species has been reported to almost instantaneously emit a bright blue luminous cloud in the sea when stimulated with artificial light. This method of light production, consisting of the ejection of luciferin and luciferase into sea water and the color of light produced by this species are similar to that of *Cypridina hilgendorfii*. In captivity, this species emitted apparent spontaneous flashes of light, whose duration was approximately 1.5 seconds, with an apparent latency of 500-800 milliseconds. Further, the luminescence of this species is due to a first-order reaction and is similar to that of *Cypridina hilgendorfii*. The luciferins and luciferases of both these species have been reported to cross-react to give light. The luminescence of *Cypridina serrata*, like *Cypridina hilgendorfii*, is oxygen-dependent. Furthermore, *Cypridin serrata* luciferin is similar to *Cypridina hilgendorfii* luciferin when compared by paper chromatography [112].

Vargula antarctica

Image not available

This species has been found distributed in the continental subregion of the Antarctic region. It has medial bristles and its rostrum bristles, however, are rather sparse. In most cases, bifurcated bristles have been observed.

Bioluminescence: It has been reported to emit a bright blue luminescent trail [109].

Vargula hilgendorfii

It is an indigenous species (Fig. **2**) in the coastal waters of Japan, and the individuals that are commonly called as "sea-fireflies" inhabiting sandy or muddy near-shore substrates. Its adult carapace length varies from 3 to 4 mm.

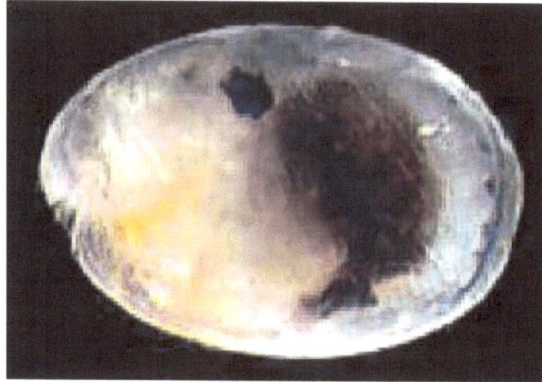

Fig. (2). *Vargula hilgendorfii.* Image credit: Ohmiya Yoshihiro, (Reproduced with permission)

Bioluminescence: The light organ, located on the upper lip (labrum) of this species contains two types of secretory cells. Each cell type presumably contributes one of the reactants necessary for its light production [110]. All individuals of both sexes, from juveniles to adults bear the yellow luminescent organ. This tiny organism has also been reported to produce copious bluish luminescence (Fig. **3**) [113] when disturbed. It is also interesting to know that if these organisms were quickly dried in the sun and kept desiccated, their luminescence would return by adding water. Further, if the dried ostracods were kept free of moisture with a desiccant, their luminescence would be preserved for indefinite periods, say decades.

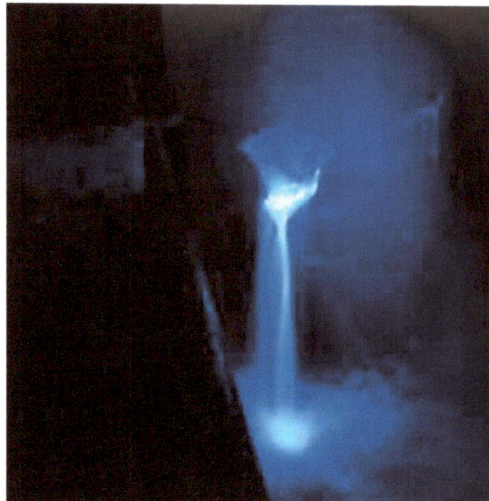

Fig. (3). Bluish luminescence of sea-firefly, Vargula hilgendorfii. Image credit: Ohmiya Yoshihiro (Reproduced with permission)

Vargula morini

Image not available

Vargula morini, a new species has been reported to occur in the Caribbean Sea reef habitats including shallow fore-reefs at depths of 3–4m. Carapace of the animals is oval with deep incisur and a protruding keel. Margins of carapace are rounded, and postero-dorsal margin is sloped more steeply than antero-dorsal margin. A little dark reddish pigmentation is seen below the eye. Gut is army-green to light brown. Eyes are maroon-black, and the curved rectangular light organ is light yellow. It has a size of about 2 mm.

Bioluminescence: The typical luminescent display of this species is a very rapid downward display consisting of about 10–20 flashes per train [114].

Vargula norvegica

Image not available

It is found distributed in the west coast of Norway at depths of 100-200 m. Dorsal and ventral margins of carapace are evenly arched. Posterior margin is rounded. Carapace is smooth without lateral processes. Bristles are seen along the inner margin and anteroventral list. It attains a maximum length of 3.7 mm.

Bioluminescence: This species luminesced in response to mechanical stimulation (manual agitation) and was also observed ejecting a viscous substance that emitted a pale blue light visible to the naked eye [115, 116].

Family: Halocyprididae

Conchoecia spinifera

Image not available

Conchoecia spinifera is found distributed in Atlantic at mesopelagic to epipelagic zones of Atlantic, Indian and Pacific Oceans. Shell is elongated and is about twice as long as high. It is thin and transparent, with a spine at the postero-dorsal corner of the right valve. maximum length of the shell is 2.2 mm.

Bioluminescence: In some males of this species, the ventral margins and the shoulder vaults appear darkly opaque due to the accumulated secretions in the bioluminescent glands lining these margins [117]. Conchoecia bioluminescence may either be excreted or retained within the glands and this species may be responsible for intense luminescence [38].

Conchoecia subarcuata

Image not available

This luminescent species is found distributed in all oceans and in the Atlantic, Indian and Pacific Oceans at depths of 200-500m. Posterior margin of the female shell is rounded to the ventral margin and its ventral margin is concave . Height is approximately half the length. Length of the shell is up to 2.2 mm.

Conchoecissa imbricata *(= Conchoecia Imbricata)*

Image credit: Angel, Martin, WoRMS

This luminescent species (Fig. **4**) is found distributed in Atlantic, Pacific and Indian Oceans. Its postero-dorsal corners of the shell are produced into sharp points considerably longer on the left than on the right. At the posteroventral corners, large processes are found. Maximum size of the shell is 3.5 mm.

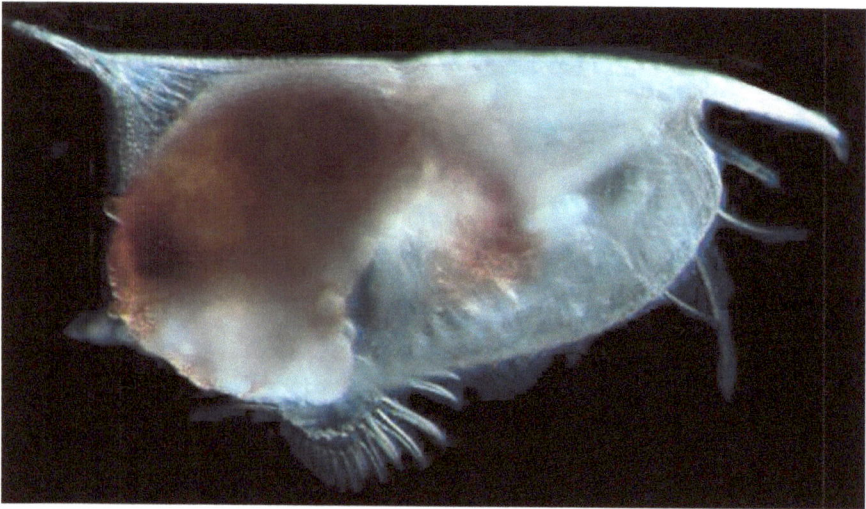

Fig. (4). *Conchoecissa imbricata.*

Conchoecia mega

No reports on its description and image

Bioluminescence: It is a bioluminescent species [118].

Discoconchoecia elegans *(= Conchoecia elegans)*

Image credit: Hopcroft, Russell, WoRMS

It is a shallow to deep water species (Fig. **5**) and is found distributed inall oceans including Atlantic where it is more common. Carapace of male tapers anteriorly and shoulder vaults are well developed in both sexes. Postero-dorsal corner of right valve carries a small spine. Carapace sculpturing is absent.

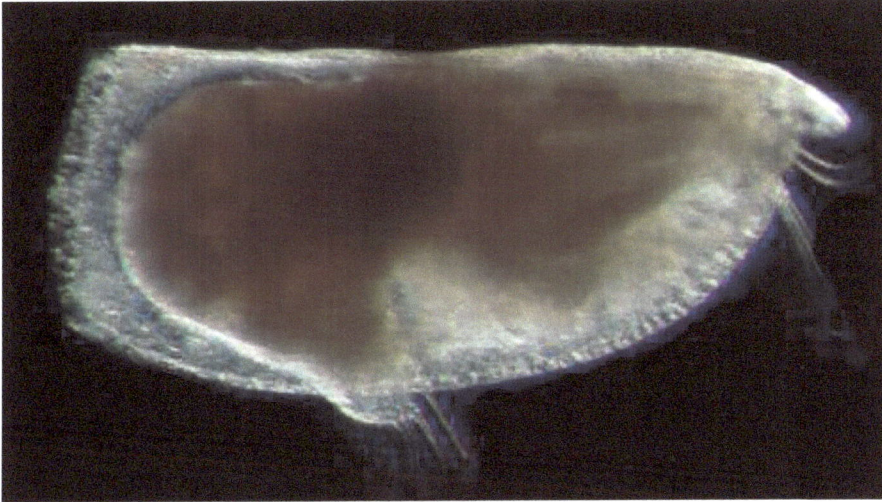

Fig. (5). *Discoconchoecia elegans.*

Bioluminescence: In response to mechanical stimulation, this species has been reported to produce quick repetitive flashes at 7 flashes per second in trains of from 2 to 120 flashes per display [38]. The quantity of photons per flash during bioluminescence is 1×10^{10} [119].

Discoconchoecia pseudodiscophora (= *Conchoecia pseudodiscophora)*

Image credit Dr. Yuichi Oba, Ph.D. (Reproduced with permission)

It (Fig. **6 - 8**) is found distributed at about 500 m depth and is endemic to the North Pacific . It is common in Okhotsk Sea and Japan Sea. Shell length of adult males and females are 1 and 1.3 mm respectively.

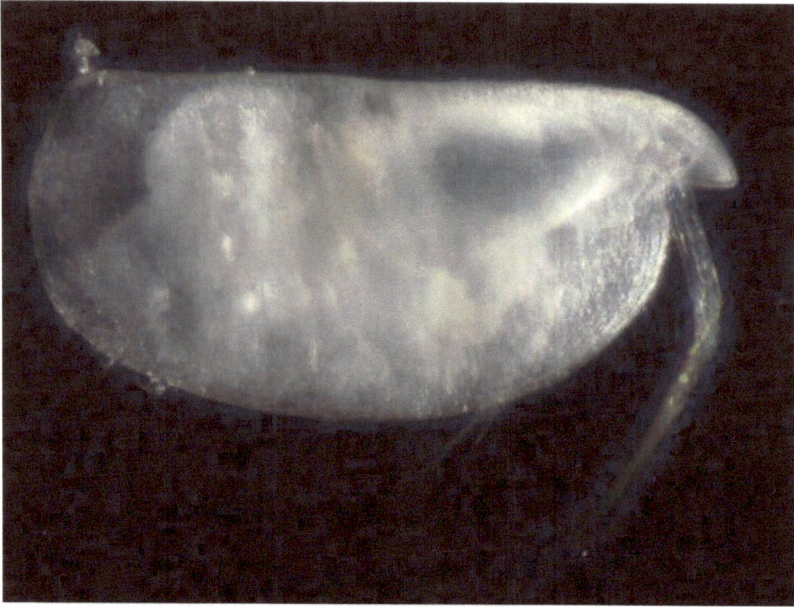

Fig. (6). *Discoconchoecia pseudodiscophora.* Image credit Dr. Yuichi Oba, Ph.D.

Fig. (7). Luminescent form.

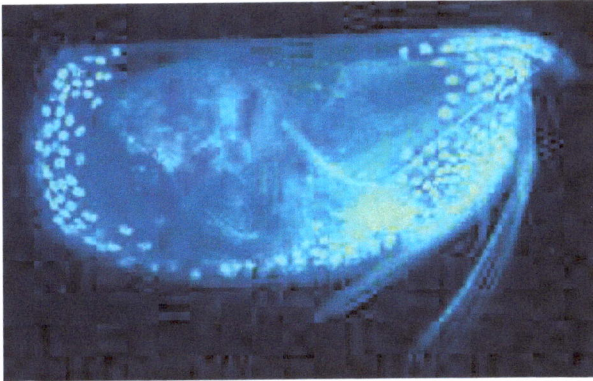

Fig. (8). Luminescence Image credit Dr. Yuichi Oba, Ph.D. (Reproduced with permission).

Bioluminescence: In this species, the luminescence (lambda(max)=463 nm) is produced by a luciferin-luciferase reaction, and its luciferin has been identified as coelenterazine. Coelenterazine, coelenteramide, and coelenteramine from this species have been quantified. The reaction between homogenates of this species and its synthetic coelenterazine showed luminescence activity which suggests that a coelenterazine-type luciferase is present [120].

Orthoconchoecia secernenda *(= Conchoecia secernenda)*

Image not available

It is a fairly massive species. Shell is opaque and grey, and pale striations may be seen starting back from rostrum . Shoulder vaults are but smoothly rounded. In some males, posterior margin is convexly rounded and the postero-ventral corner is rounded to the ventral margin. Further, spines at the posterodorsal corner are longer and slimmer in *Conchoecia bispinosa*.

Bioluminescence: This luminescent species produced light during mechanical stimulation and its bioluminescent potential was similar in magnitude to that reported for other bioluminescent plankton such as larger dinoflagellates, copepods and larvaceans [121].

BIOLUMINESCENT MARINE COPEPODS

Among the zooplankton fauna, the copepods "the insects of the sea" are represented with about 2500 species in the oceans worldwide. They are widely distributed in the northeastern Atlantic, Antarctic and western parts of the Indian Ocean (Table **1**) and are known to provide functionally important links in the aquatic food chain [122]. However, luminescence has been reported only in about

60 species belonging to the Order Calaoida and Order Cycopoida (Table **2**). The families Metridinidae, Heterorhabdidae, Lucicutiidae, and Augaptilidae of the order Calanoida are well known for luminescence. Among the cyclopoid copepods, *Oncaea spp.* are only notable for their luminescence [123]. The luminous organs (glands) which are of greenish-yellow or yellow in live specimens are located on their appendages, such as the tail or swimming legs. The number and location of these luminous glands, however, differ considerably among the families of copepods. With physical stimulus, these copepods discharge a luminous secretion; and they are also capable of producing light internally [61]. The luminescence is used by these copepods to detract their predators [13, 38]. The larval stages of copepods such as *Centropages furcatus, Paracalanus Indiens, Acrocalanus longicornis, Corycaeus (Corycaeus) speciosus* and *Corycaeus (Onychocorycaeus) latus* are also bioluminescent [124]. In the luminescent arthropods, colenterizine (as luciferin) is present but there is no Ca^2 activated photoprotein [27].

Copepod luciferases: family of small secretory proteins of 18.4–24.3 kDa, including a signal peptide—are responsible for bright secreted bioluminescence of some marine copepods. The copepod luciferases use coelenterazine as a substrate to produce blue light in a simple oxidation reaction without any additional cofactors [32].

Coelenterazine containing copepod species: *Acartia claui, Calanus sp., Centrophagus typicus, Euaugaptilus magns, Euaugaptilus laticeps, Euaugaptilus peodiosus, Gaussia princeps, Hemirhabdus grimaldi, Heterorhabdus tanneri, Heterostylites major, Lucicutia ovaliformis, Megacalanus princeps, Metridia longa, Metridia pacifica, Metridia longa, Metridia okhotensis, Oithona helgolandia, Pareuchaete spp., Pleuromamm abdominalis, Pleuromamma piseki,* and *Temora logicornis* [72, 85, 123].

Emission Maxima in Luminescent Marine Planktonic Copepods: The emission maxima (λmax) of the different species of luminescent marine planktonic copepods have been reported to vary from 430 to 517 nm (Table **5**).

Table 3. Global distribution and habitat of the luminescent marine copepods [13].

Species	Global distribution	Habitat
Centraugaptilus horridus	Indian Ocean	Shallow and deep , 0-2070m
Centraugaptilus rattrayi	Cosmopolitan except Arctic and North Pacific	Shallow and deep, 0-1000m
Disseta palumbii	W.Pacific	Shallow and deep, 50-500m

(Table 3) cont.....

Euaugaptilus magnus	Cosmopolitan, except for Arctic, NE Pacific, California and Red Sea	Surface, deep, 0-1570m
Euaugaptilus parabullifer	NW Pacif., NE Japan	Deep, 2000-3000m
Euaugaptilus squamatus	Pacific, N.Atlantic	Shallow and deep, 0-3000m
Gaussia princeps	Worldwide	Deep, Mesopelagic ,200-1000m
Hemirhabdus grimaldii	Atlantic, Indian and Pacific Oceans	Deep, 500-1500m
Heterorhabdus norvegicus	Pacific, N.Atlantic	Shallow and deep, 0-8000m
Heterorhabdus papilliger	Western Pacific: China, Taiwan and Kermadec	Deep, 200-1000m
Heterorhabdus tanneri	Pacific	Shallow and deep, 65-3000m
Heterostylites longicornis	Northwest Pacific: China	Shallow and deep , 0-300m
Heterostylites major	Atlantic, Indian and Pacific Oceans	Deep, 500 - 1000 m
Lucicutia bicornuta	Atlantic, Indian and Pacific Oceans	Shallow and deep , 300-4000m
Lucicutia flavicornis	Cosmopolitan	Shallow and deep, 0-500m
Lucicutia gaussae	Atlantic, Indian and Pacific Oceans	Shallow and deep, 0-2000m
Lucicutia grandis	Atlantic, Indian and Pacific Oceans	Shallow and deep , 0-6000m
Lucicutia wolfendeni	Atlantic, Indian and Pacific Oceans	Shallow and deep , 0-4000m
Metridia gerlachei	Antarctic and sub-Antarctic; southern Atlantic, Pacific, and Indian Oceans	Shallow and deep, 0-1000
Metridia longa	Arctic, N.Atlantic, Pacific	Shallow, 50-200m
Metridia lucens lucens	Arctic, Southeast Pacific and Antarctic Indian Ocean	Shallow and deep, 0-500m
Metridia pacifica	Subarctic Pacific	Shallow and deep,100- 800m
Metridia okhotensis	Pacific	Shallow and deep, 0-2000m
Metridia princeps	Antarctic	Antarctic Shallow and deep, 50-3000m
Paraheterorhabdus robustus	Atlantic and Pacific Oceans	Deep, 1000 – 1900m
Pleuromamma abdominalis abdominalis	Atlantic (Gulf of St. Lawrence)	Shallow, 0 - 200 m
Pleuromamma borealis	Atlantic, Indian and Pacific Oceans ; Arctic	Shallow and deep, 0-700m
Pleuromamma piseki	Atlantic, Indian and Pacific Oceans	Shallow, 0-170m
Pleuromamma robusta robusta	Atlantic, Pacific, Arctic	Shallow and deep, 0-500m
Pleuromamma scutellata	Pacific	Shallow and deep, 250-500m
Pleuromamma xiphias	Cosmopolitan except Arctic	Shallow and deep, 0-950m
Triconia conifera	Arctic, Antarctic, Pacific	Surface, 30-170m

Table 4. Bioluminescent marine copepods [13].

Class	Order	Family	Species
Copepoda	Calanoida	Augaptilidae	*Centraugaptilus horridus, C. rattrayi, C.cucullatus, Euaugaptilus bullifer, E. farrani, E. filigerus, E. grandicornis, .E laticeps, E.magnus, E. nodifrons, E. perodiosus, E. rectus, E. squamatus, E. truncates, E, vicinus, Haloptilus longicirrus, H.acutilobus, Pachyptilus eurygnathus*
-	-	Heterorhabdidae	*Disseta palumboi, Hemirhabdus grimaldii, H. latus, H. longicornis, H. norvegicus, H. papilliger, H. robustus, H. spinifrons*
-	-	Lucicutiidae	*Lucicutia aurita, L. clausi, L. flavicornis, L. gemina, L. grandis, L. magna, L. ovalis, L. sarsi, L. wolfendeni*
-	-	Metridinidae	*Gaussia princeps, Metridia gerlachei, M. longa, M. lucens, M macrura, M. pacifica, M. princeps, Pleuromamma abdominalis, P. borealis, P.gracilis, P. indica, P.piseki, P.quadrungulata, P. robusta, P. xiphias*
-	Cyclopoida	Corycaeidae	*Corycaeus latus, C. speciosus*
-	-	Oncaeidae	*Oncaea conifera*

Table 5. Emission maxima (λmax) of luminous marine copepods.

Species	max	Ref
Chiridius poppei	517 nm	[125]
Euaugaptilus magnus	480nm	[10]
Gaussia princeps	485nm	[65]
-	480,485nm	[10]
-	479, 489 ; 483-489 nm	[13]
Lucicutia ovlis	493nm	[123]
Metridi longa	485nm	[65]
Metridia lucens	482nm	[10]
Metridia okhotensis	493nm	[123]
Oncaea conifera	430,435nm	[10]
Pleuromamma abdominalis	486 ,465nm	[13]
-	493, 490nm	[123]
Pleuromamma borealis	470(485); 480.485,490nm	[10]
Pleuromamma xiphias	492, 472nm	[13]

LUMINESCENT SPECIES OF MARINE COPEPODS

Luminescent copepods have been reported to discharge a luminous secretion when stimulated electrically or mechanically. Several species of copepods species possess a number of luminescent glands. *Metridia lucens* for example, has glands on its head, middle thorax and urosome. In live animals, luminescent material may be discharged simultaneously from all these glands. Several patches of glowing secretion of these copepods may be seen in the water. It was also observed that düring repeated stimulation, flashes may become weak and some glands may even cease discharging before others. In fact a luminous response is said to be a complex event consisting of several luminous discharges with varying temporal and spatial characteristics. Further, the glands may also luminesce within the animal [61].

Order: Calanoida

Centraugaptilus horridus

Image not available

Luminous organs/Bioluminescence: In the species of *Centraugaptilus*, large yellow-green luminous glands are present on the distal two segments of the exopods of the swimming legs [61].

Centraugaptilus rattrayi

Image not available

Luminous organs/Bioluminescence: In the species of Centraugaptilus, large yellow-green luminous glands are present on the distal two segments of the exopods of the swimming legs [61].

Centropages typicus

Image credit: Daniel J. Drew, Wikimedia

Luminous organs/Bioluminescence: Not reported in this species (Fig. **9**).

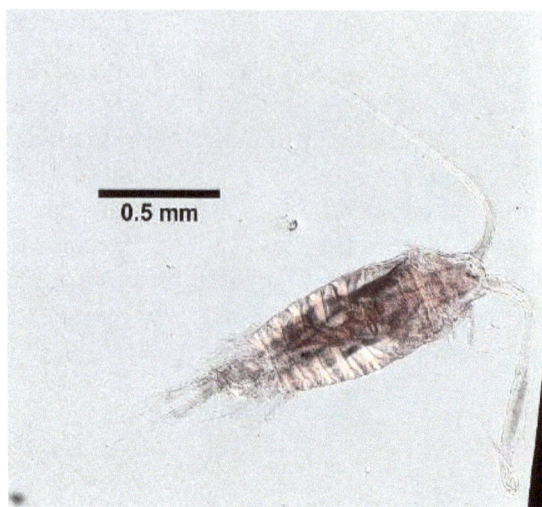

Fig. (9). *Centropages typicus.*

Chiridius poppei

Image not available

Luminous organs/Bioluminescence: In this luminescent species, a new class of protein called CpYGFP has been isolated from this species and this protein had excitation and emission maxima at 507 and 517 nm, respectively. The usefulness of this protein as a reporter in the subcellular localization of actin has also been demonstrated [125]. Actin is used presently in the treatment of various allergic conditions

Euaugaptilus laticeps

Image not available

Luminous organs/Bioluminescence: In all the species of *Eugaptilus,* large yellow-green luminous glands are present on the distal two segments of the exopods of the swimming legs [61]. Further, in these species, more than 90% of the luciferase is found in their legs, with the luminous cells, but over 40% of the coelenterazine is found in their bodies [27].

Euaugaptilus magnus

Image not available

Luminous organs/Bioluminescence: It possesses luminous glands, the four secretory cells of which are innervated by the same nerve. With electrical stimulation, its luminescent intensity was found to be 0.66 x 10^{-5} µW/cm^2 at 15 cm [61].

All the luminescent *Euaugaptilus* species have been reported to possess more than 90% of the luciferase in their luminous cells of legs and about 40% of the coelenterazine (coeleterate-type luciferin) in their bodies [27].

Gaussia princeps

Image credit: Flickr

Luminous organs/Bioluminescence: This species (Fig. **10**) has the maximum number (70) of bioluminescent glands, or bioluminescent sites. These glands with coeleterate-type luciferin are present in the head, mandibular palps, urosome and furca of the animal. During rapid swimming, this species has been reported to produce bright, blue luminescent displays from the luminous glands, body parts and underlying body pores. The light so emitted is a peak intensity for 1-3 sec [85, 13]. This species also possesses a novel luciferase which catalyzes the oxidation of its coelenterazine to produce the light [123].

Fig. (10). *Gaussia princeps.*

Heterorhabdus norvegicus

Image credit: Fisheries and Oceans Canada, J.M. Spry WoRMS

Luminous organs/Bioluminescence: In this species (Fig. **11**), several pairs of luminous glands are seen in its head region, thorax anal segment and caudal rami; thoracic appendages and distal segments of the swimming legs; and first three segments of the first antennae, exopodite of the mandibles and second maxillae, second antennae and first maxillae [61].

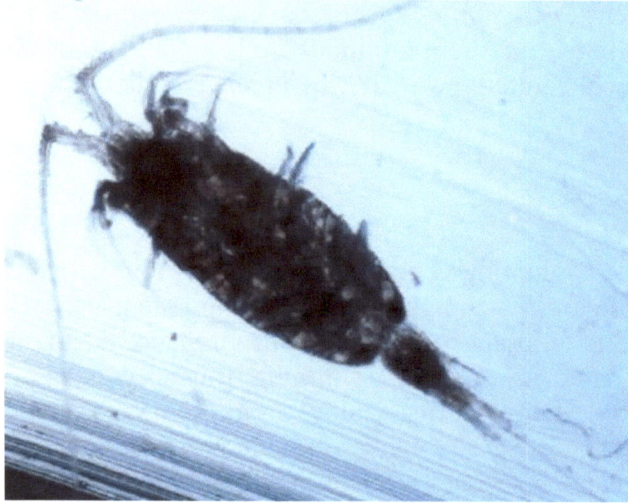

Fig. (11). *Heterorhabdus norvegicus.*

Heterorhabdus papilliger

Image not available

Luminous organs/Bioluminescence: In this species, 36 pairs of luminous glands are seen in its head region, thorax anal segment and caudal rami; thoracic appendages and distal segments of the swimming legs; and first three segments of the first antennae, exopodite of the mandibles and second maxillae, second antennae and first maxillae [61].

Heterostylites longicornis

Image credit: Fisheries and Oceans Canada, Moira Galbraith WoRMS

Luminous organs/Bioluminescence: It (Fig. **12**) possesses luminous glands . With electrical stimulation this species has been reported to flash for a duration of 3 sec. and its luminescent intensity was found to be 0.12 x 10^{-5} μW/cm^2 at 15 cm [61]

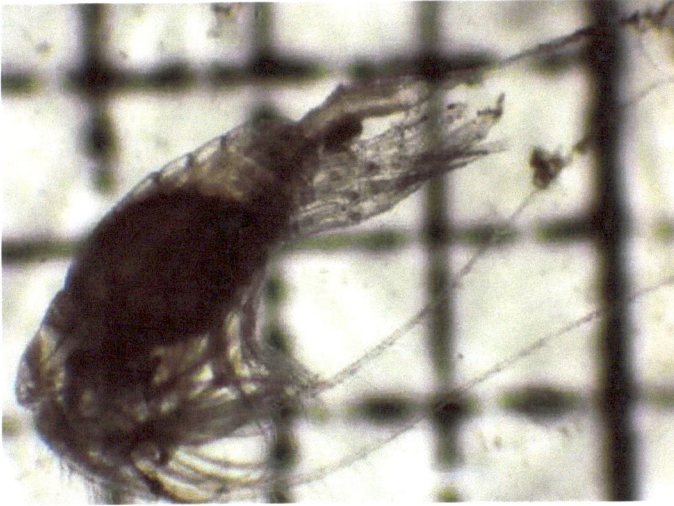

Fig. (12). *Heterostylites longicornis.*

Lucicutia flavicornis

Image credit: Dolan, John WoRMS

Luminous organs/Bioluminescence: In this species (Fig. **13**) several pairs of large luminous glands are believed to be present in the anterior ventral margin of the first thoracic segment [61].

Fig. (13). *Lucicutia flavicornis.*

Lucicutia grandis

Image credit: Karen Wishner (Reproduced with permission)

Luminous organs/Bioluminescence: In this species (Fig. **14**), 10 large luminous glands are present. With electrical stimulation this species has been reported to flash for a duration of 0.2 - 2 sec. and its luminescent intensity was found to be $0.12 - 0.70 \times 10^{-5}$ µW/cm2 at 15 cm [61].

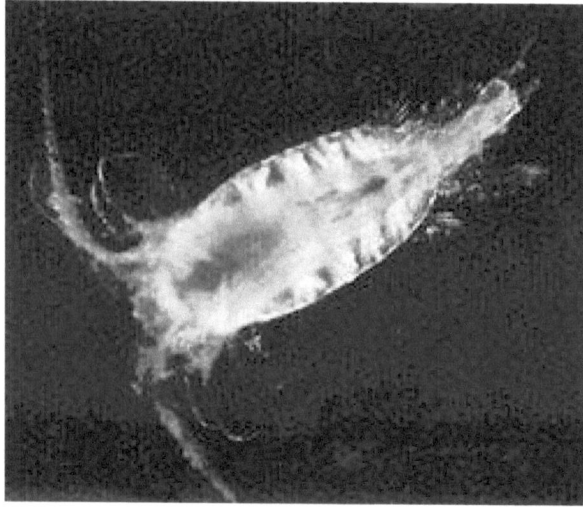

Fig. (14). *Lucicutia grandis.*

Lucicutia wolfendeni

Image not available

Luminous organs/Bioluminescence: The luminous glands of this genus are fluorescent [61].

Metridia gerlachei

Image not available

Luminous organs/Bioluminescence: A pair of luminous glands located on the caudal rami of this species have been reported to emit luminescent trail [113].

Metridia longa

Image credit: Arctic Ocean Diversity

Luminous organs/Bioluminescence: This species (Fig. **15**) has been reported to eject copious amounts of luminous materials from its while swimming vigorously [126]. In this species, a pair of luminous glands opens just dorsal to the middle caudal bristle on each caudal ramus . Similarly another of glands opens on the lateral posterior corner of the anal segments. However, it has no glands on its thorax. Further, the glands are clustered near the midline of its head. The total number of glands in this species is believed to be 13. With electrical stimulation this species has been reported to flash for a duration of 0.5 -6.3 sec. and its luminescent intensity was found to be 0.07- 1.5 x 10^{-5} $\mu W/cm^2$ at 15 cm [61].

Fig. (15). *Metridia longa.*

Each of the two cell types of this species produced different components needed for light emission *viz.* 'luciferin' and 'luciferase' respectively [127]. They also hypothesized that when these organisms are stimulated, each cell secretes its contents, generating light as the materials combined in the surrounding seawater where they are ejected (Larionova *et al.*, 2020) . The nauplius and adult of this species produced photons per flash as $2x10^9$ and $2x10^{11}$ respectively [129]. A distinct Luciferase gene (MLuc) has been isolated from this species [123].

Metridia lucens

Image not available

Luminous organs/Bioluminescence: In this species, a pair of luminous glands opens just dorsal to the middle caudal bristle on each caudal ramus . Similarly another of glands opens on the lateral posterior corner of the anal segments. Its conspicuous luminous glands have been reported to open laterally on the its second thoracic segment. In this species, its 10 luminous glands are found arranged in a definite pattern *viz.* three in a row along the anterior edge of the carapace just dorsal to the rostrum; and the remaining one gland is centrally located. The total number of glands in this species is believed to be 13. In this species it is worth noting that in its bright bioluminescence, coelenterazine luciferin luciferin production and luciferase transcription are involved. With electrical stimulation, this species has been reported to flash for a duration of 0.1-5.5 sec. and its luminescent intensity was found to be 0.02-0.77 x 10^{-5} µW/cm^2 at 15 cm [61]. Its photons per flash was to the tune of 4×10^{10} [119].

Metridia pacifica

Image not available

Luminous organs/Bioluminescence: The distinct Luciferase genes (MpLuc1 and MpLuc2) have been isolated from this species [123]. In this species, putative luminous glands are seen on its head, legs and abdomen adjacent to the head.

Metridia princeps

Image not available

Luminous organs/Bioluminescence: In this species, large luminous glands are present in its anal segment; single glands on either side of its thoracic segments one and two; a single gland or small cluster at the most anterior margin of the head; two fluorescent spots on either side of the midline at the shoulder between the dorsal and lateral body wall; and conspicuous glands in its basipods of the first and second swimming legs. With electrical stimulation, this species has been reported to flash for a duration of 1.6^{-7} sec. and its luminescent intensity was found to be 0.17- 9.4 x 10^{-5} µW/cm^2 at 15cm [61].

Paraheterorhabdus robustus *(= Heterorhabdus robustus)*

Image not available

Luminous organs/Bioluminescence: It possesses luminous glands . With electrical stimulation this species has been reported to flash for a duration of 0.2 - 2.5 sec. and its luminescent intensity was found to be 0.01- 0.05 x 10^{-5} µW/cm^2 at 15cm [61].

Pleuromamma abdominalis abdominalis

Image credit: Flickriver

Luminous organs/Bioluminescence: The sub-cuticular cells present in the skin glands of this species (Fig. **16**) have been reported to be responsible for the production of its luminescence. Further, its greenish-yellow secretion is also luminescent [127].

Jose Manuel Gutiérrez Salcedo (2017)

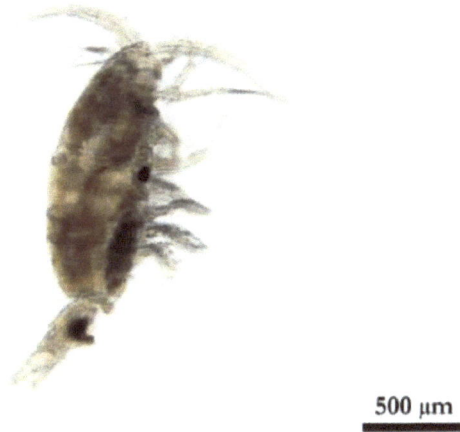

500 μm

Fig. (16). *Pleuromamma abdominalis abdominalis.*

Pleuromamma robusta

Image not available

Luminous organs/Bioluminescence: In this species. Three luminous glands are seen just dorsal to the rostrum and another two glands in the posterior and lateral region of the head at the level of its mandible. Double glands have also been reported on its 2nd thoracic segment and paired glands are seen on its anal segment and caudal rami [61].

Pleuromamma xiphias

Image credit: Dr. Amy Maas (Reproduced with permission)

Luminous organs/Bioluminescence: It (Fig. **17**) possesses two luminous glands on the anterior portion of its head on either side of the crest near its base. With electrical stimulation this species has been reported to flash for a duration of 0.3 - 6.1 sec. and its luminescent intensity was found to be $0.34 - 1.29 \times 10^{-5}$ μW/cm^2 at 15cm [61].

Fig. (17). *Pleuromamma xiphias.*

Order: Cyclopida

Corycaeus spciosus

Image credit: Uribe-Palomino, Julian, WoRMS

Luminous organs/Bioluminescence: Not reported in this species (Fig. **18**).

Fig. (18). *Corycaeus spciosus.*

Triconia conifera (= *Oncaea conifera*)

Image not available

Luminous organs/Bioluminescence: It has a large number of epidermal, unicellular luminous glands which are distributed mainly over the dorsal cephalosome and urosome. In this species, bioluminescence is produced in the form of short (80 to 200 ms duration) flashes from within each gland and there is no visible secretory component. Each luminous gland seems to open to the exterior by a simple valved pore. Intact copepods have been reported to produce several hundred flashes before its luminescent system is exhausted. Individual flashes bore a maximum flux of 7.5 X 10^{10} quanta s^{-1} [128].

BIOLUMINESCENT MARINE AMPHIPODS

Among the planktonic crustaceans, the amphipods have rather limited luminescent species amounting to only 10 (Table **6**)

Table 6. Bioluminescent marine amphipods [13].

Class	Order	Family	Species
Malacostraca	Amphipoda	Cyphocarididae	*Cyphocaris faurei*
-	-	Lanceolidae	*Megalanceola terranovae*
-	-	Scinidae	*Scina borealis, S. crassicornis, S. marginate, S. rattrayi, S. submarginata*
-	-	Pronoidae	*Parapronoë crustulum*
-	-	Oxycephalidae	*Streetsia nyctiphanes, S. porcella*

Emission Maxima in Luminescent Marine Planktonic Amphipods: The emission maxima (λmax) of the different species of luminescent marine planktonic amphipods have been reported to vary from 415 to 475 nm (Table **2**).

Luminescent Planktonic Marine Amphipods

Cyphocaris faurei

Image not available

Depth of occurrence and Distribution: Bathypelagic; Tropical; Western Central Pacific: Indonesia and New Caledonia.

Luminous organs/Bioluminescence: In this species, bioluminescence appeared as a secretion through integumentary pores on its telson and uropods, and as a glow from a single location on its cephalothorax. Its emission spectrum has been reported to be bimodal or unimodal, with distinct blue-green and orange peaks [129].

Phtisica marina

Image credit: Flickr

Depth of occurrence and Distribution: 0 - 1470 m; Subtropical; Atlantic Ocean, Eastern Central Pacific and the Mediterranean: Europe and USA.

Luminous organs/Bioluminescence: This luminous species (Fig. **19**) has been reported to live on Obelia colonies [72].

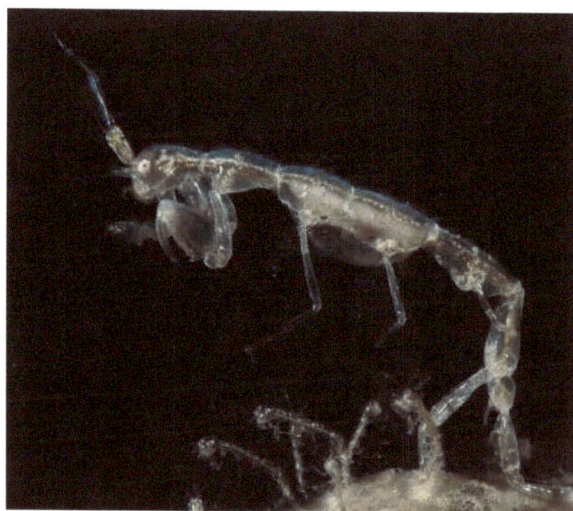

Fig. (**19**). *Phtisica marina.*

Scina borealis

Image not available

Depth of occurrence and Distribution: 50 - 3000 m; Subtropical to polar; Western Pacific, Western Central Atlantic and the Arctic.

Luminous organs/Bioluminescence: The bioluminescence of this species is internal and it is of significantly shorter duration and lower quantum emission, with a unimodal, blue-green emission spectrum [129].

Scina crassicornis

Image not available

Depth of occurrence and Distribution: 0 - 500 m; Subtropical; Western Pacific: New Caledonia; South China Sea.

Luminous organs/Bioluminescence: The bioluminescence of this species is internal and it is of significantly shorter duration and lower quantum emission, with a unimodal, bluegreen emission spectrum [129].

Scina rattrayi

Image credit: SeaLifeBase CC

Depth of occurrence and Distribution: This luminescent species (Fig. **20**) occurs at depths up to 4600 m in Subtropical; Northwest Pacific and Antarctic Atlantic.

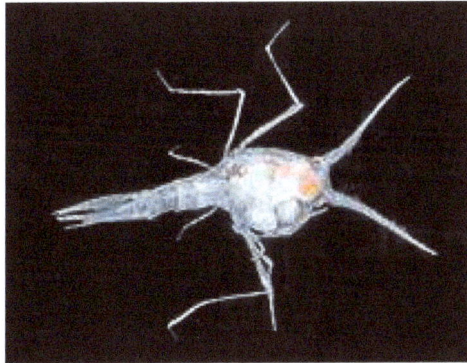

Fig. (20). *Scina rattrayi.*

Luminous organs/Bioluminescence: Not reported

BIOLUMINESCENT MARINE MYSIDS

The planktonic, marine mysids have very poor luminescent representatives which are hardly less than 5 (Table **8**).

Table 7. Emission maxima (λmax) of luminous marine amphipods.

Species	λ max	Ref
Scina marginata	435nm	[10]
Scina rattrayi	439nm	[13,42]
Parapronoë crustulum	470(410); 475(405); 475(415)nm	[10]

Table 8. Bioluminescent marine mysids [13]

Class	Order	Family	Species
Malacostraca	Mysidacea	Lophogastridae	*Gnathophausia longispina, G.zoea, Neognathophausia gigas, N.ingens*

Emission Maxima in Luminescent Marine Planktonic Mysids: The emission maxima (λmax) of the different species of luminescent marine planktonic mysids have been reported to vary from 481 to 484 nm (Table **9**).

Table 9. Emission maxima (λmax) of luminous marine mysids.

Species	λ max	Ref
Neognathophausia ingens	481nm	[85]
-	484nm	[13]

Luminescent Marine Mysids

Gnathophausia longispina

Image not available

Depth of occurrence and Distribution: 300 - 1650 m; Tropical; Indo-Pacific

Luminous organs/Bioluminescence: The luminescence of Gnathophausia spp. is due to their coelenterate-type luciferin/luciferase [85]. Further, these species have also been reported to eject a luminous secretion (luminous cloud) into the water).

Gnathophausia zoea

Image credit: Fisheries and Oceans Canada, WoRMS

Depth of occurrence and Distribution: 400 - 6050 m; Subtropical; Cosmopolitan

Luminous organs/Bioluminescence: This species (Fig. **21**) has been reported to show extremely low reflectances at blue-green wavelengths (<0.5%) and much higher reflectances at longer wavelengths [130].

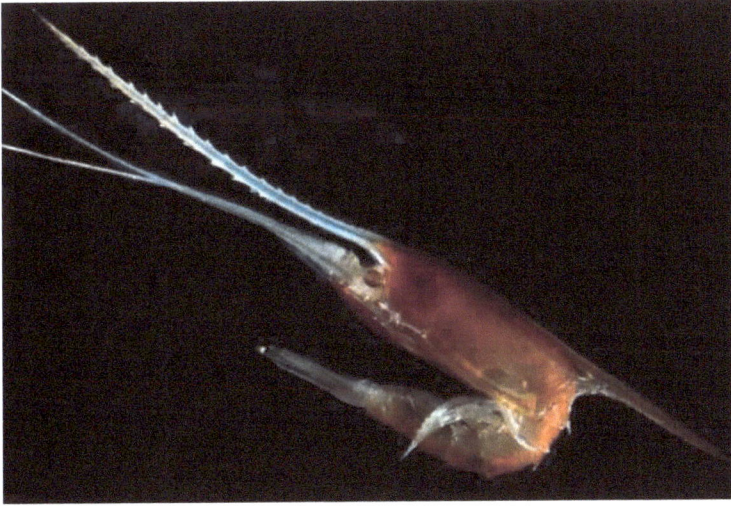

Fig. (21). *Gnathophausia zoea.*

Neognathophausia gigas *(= Gnathophausia gigas)*

Image not available

Depth of occurrence and Distribution: 600-4400 m; Cosmopolitan in tropical and temperate waters

Luminous organs/Bioluminescence: This species has a luminous gland on its second maxillae from which it ejects a brilliantly luminescent cloud into the seawater when it is disturbed [131].

Neognathophausia ingens *(= Gnathophausia ingens)*

Image not available

Depth of occurrence and Distribution: 500- 4000 m; Worldwide in tropical and temperate seas.

Luminous organs/Bioluminescence: This giant red mysid has a luminous gland on its second maxillae from which it ejects a brilliantly luminescent cloud into the seawater when it is disturbed [85].

BIOLUMINESCENT MARINE EUPHAUSIIDS

Euphausiids (in Greek, "true light emitting") or krills are small holoplanktonic shrimp-like crustaceans. With the exception of *Bentheuphausia ambylops* and

Thysanopoda minyops light organs (photophores) are present in all species of euphausiids in the family Euphausiidae. A total of 36 species of euphaussids have been reported to be luminescent [13]. In these euphausiids, the light organs are located on various parts of the individual krill's body *viz.* one pair at the eyestalk, another pair are on the hips of the second and seventh thoracopods, and singular organs on the four pleonsternites. These light organs have been reported to emit a yellow-green light periodically, for up to 2–3 s. They are highly developed and can be compared with a flashlight. In each light organ, there is a concave reflector in its back and a lens in the front that guide the light produced. The whole light organ can be rotated by muscles, which can direct the light to even a specific area. For light production, these euphausiids have been found to utilise a tetrapyrole luciferin and are not known to contain coelenterazine. The krill's bioluminescent organs have also been reported to contain several fluorescent substances and the major component has a maximum fluorescence at an excitation of 355 nm and emission of 510 nm [72]. The emission spectra of homogenates of many species of euphausiids show maxima in a narrow range from 467 to 473 nm [132].

Emission Maxima in Luminescent Marine Planktonic Euphausiids: The emission maxima (λmax) of the different species of luminescent marine planktonic euphausiids have been reported to vary from 453 to 540 nm (Table **10**).

Table 10. Emission maxima (λmax) of luminous marine euphausiids.

Species	λmax	Ref
Euphausia americana	465,470nm	[10]
-	467-473nm	[13]
Euphausia brevis	470nm	[10]
-	467-473nm	[13]
Euphausia frigida	470nm	[10]
Euphausia gibboides	470nm	[10]
-	467-473nm	[13]
Euphausia hemigibba	470nm	[10]
-	467-473nm	[13]
Euphausia krohnii	470,475nm	[10]
Euphausia pacifica	470nm	[42]
-	476, 520-540nm	[10]
-	476nm	[132]
Euphausia superba	470,475nm	[10]

(Table 10) cont.....

Euphausia tenera	468nm	[10]
Euphausia triacantha	475nm	[10]
Meganyctiphanes norvegica	475,476, 520-540nm	[110]
-	476nm	[132] .
Nematobrachion flexipes	453nm	[13]
Nematoscelis difficilis	483nm	[42]
Nematoscelis megalops	463nm	[132]
-	463nm	[13]
-	465,470nm	[10]
Nematoscelis microps	463nm	[13]
Nyctiphanes couchii	470 nm	[10]
-	467-473nm	[13]
Nyctiphanes simplex	467nm	[42]
Stylocheiron abbreviatum	465,470nm	[10]
Thysanoëssa gregaria	470nm	[10]
Thysanoëssa macrura	470-475nm	[10]
Thysanoëssa raschii	476, 520-540nm	[10]
-	476nm	[132]
Thysanopoda monacantha	465nm	[10]
-	467-473nm	[13]
Thysanopoda tricuspidata	465nm	[10]

Luminescent Planktonic Marine Euphausiids

Class: Malacostraca

Order: Euphausiacea

Family: Euphausiidae

Euphausia americana

Image credit: Bárbara Santos Menezes (Reproduced with permission)

Depth of occurrence and Distribution: 0 - 2500 m;, Cosmopolitan species (Fig. 22)

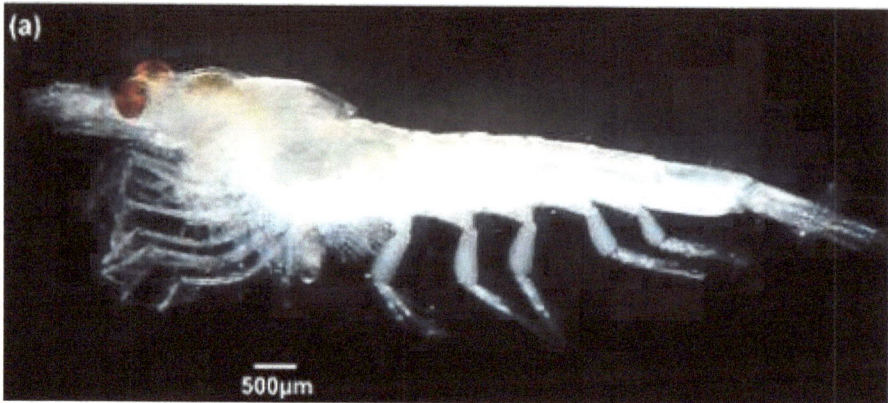

Fig. (22). *Euphausia americana.*

Diagnosis & Size: 12 mm in length; Eyes are rounded and medium in size. Rostrum is long and sharply pointed, reaching anterior limit of the eyes. There are two pairs of lateral carapace denticles and gastric region is domed.No dorsal spines are seen in abdomen. It attains a maximum length of 12 mm.

Luminous organs/Bioluminescence: Not reported

Euphausia brevis

Image not available

Depth of occurrence and Distribution: 0 - 1872 m; Tropical Pacific and Western Central Atlantic.

Diagnosis & Size: Eyes are rounded and larger in size. Rostrum is straight and acute and is reaching anterior limit of the eyes. There are two pairs of lateral carapace denticles. No dorsal spines are seen in abdomen. It attains a maximum length of 10 mm.

Luminous organs/Bioluminescence: Not reported

Euphausia vrystallorophias

Image not available

Depth of occurrence and Distribution: 100–500 m; Antarctic peninsula and continent; South Shetland Islands

Diagnosis & Size: Eyes are rounded and larger in size. Rostrum is long and sharp and is reaching front of the eyes. A gastro-hepatic transverse groove and one pair of lateral denticles are seen in abdomen. No dorsal spines are seen in abdomen. It attains a maximum length of 35 mm.

Luminous organs/Bioluminescence: The two rows of photophores located on the abdomen of this species have been reported to emit bright luminescence [113].

Euphausia eximia

Image not available

Depth of occurrence and Distribution: 0-300m; Pacific Ocean; California and Humboldt Currents and Gulf of California

Diagnosis & Size: Eyes are rounded and larger in size. Rostrum is long and sharply pointed. No spine is seen in abdominal segment. It attains a maximum length of 28 mm.

Luminous organs/Bioluminescence: Not reported

Euphausia frigida

Image credit: Siegel, Volker, WoRMS

Depth of occurrence and Distribution: Above depths of about 300 m; Polar, Southwest Atlantic

Diagnosis & Size: It (Fig. **23**) is commonly known as pygmy krill . Eyes are rounded and larger in size. Rostrum is absent and frontal plate is very short and triangular with a small acute tip . One pair of lateral denticles is seen in carapace. No dorsal spines are seen in abdomen. It attains a maximum length of 24 mm.

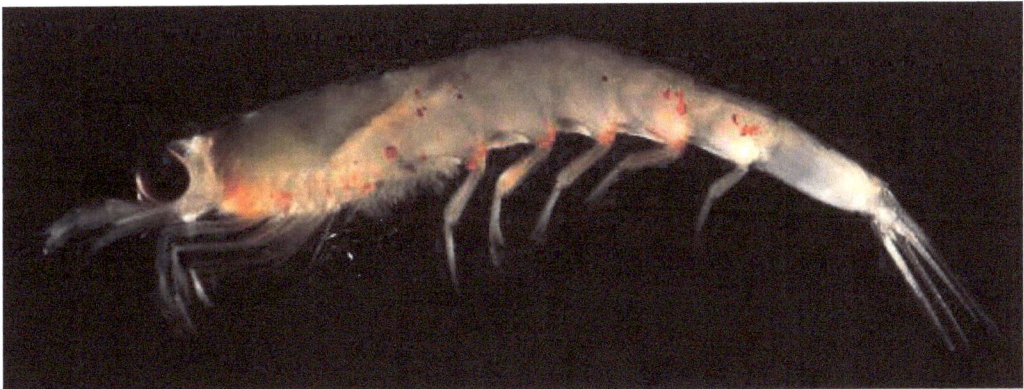

Fig. (23). *Euphausia frigida.*

Luminous organs/Bioluminescence: Not reported

Euphausia gibboides

Image not available

Depth of occurrence and Distribution: 300-400 m during the day and near the surface during the night; Pacific and Atlantic

Diagnosis & Size: Eyes are rounded and larger in size. Rostrum is long and sharply pointed . A spines is seen in 3rd abdominal segment. It attains a maximum length of 30 mm.

Luminous organs/Bioluminescence: Not reported

Euphausia hemigibba

Image not available

Depth of occurrence and Distribution: About 100 m in night and 400-550 m in daytime; Cosmopolitan

Diagnosis & Size: Eyes are rounded and small in size. Rostrum is slightly upturned and is rarely extending to the anterior margin of eye. Gastric region is a low dome. One pair of lateral denticles are seen in carapace. A mid-dorsal spine is seen in abdomen. It attains a maximum length of 17 mm.

Luminous organs/Bioluminescence: Not reported

Euphausia krohnii

Image not available

Depth of occurrence and Distribution: 400-600 m; Temperate and subtropical North Atlantic;tropical Atlantic and Mediterranean

Diagnosis & Size: Eyes are rounded and large in size. Rostrum is long, sharply pointed and is reaching the anterior limit of the eyes. Gastric region is domed. Two e pairs of lateral denticles are seen in carapace. No dorsal spines in abdomen. It attains a maximum length of 19 mm.

Luminous organs/Bioluminescence: During the larval development of this species, the ocular photophore is the first to develop and appears from stage 3 calyptotis [13].

Euphausia pacifica

Image credit: Fisheries and Oceans Canada, Moira Galbraith, WoRMS

Depth of occurrence and Distribution: 0-1000 m; North Pacific; Bering Sea to Baja California and in the Sea of Japan.

Diagnosis & Size: In this luminescent species (Fig. **24**), eyes are rounded and large in size. Rostrum is absent. No dorsal spines in abdomen. It attains a maximum length of 25 mm.

Fig. (24). *Euphausia pacifica.*

Luminous organs/Bioluminescence: In this species, a fluorescent compound F has been isolated [133]. The terminal glows of this species showed a bimodal emission spectrum with a sharp primary peak at 476 nm and a second lower peak between 520 and 540 nm [132].

Euphausia superba

Image credit: Krill666.jpg: Uwe Kils, Wikipedia

Depth of occurrence and Distribution: 0-600 m; Southern and Indian Antarctic Oceans; Antarctic Peninsula.

Diagnosis & Size: In this luminescent species (Fig. **25**), eyes are rounded and male's eye is medium in size and the female's is small. Rostrum is short and is extending to the mid-point of the eye; and rounded at the tip. Front part of carapace is produced at the anterior-lateral corners into strong, oblique projections behind each eye; and it has one pair of lateral denticles. In older males and sometimes females these denticles are either reduced or absent. Adults reach a maximum length of 65 mm. Its life span is about 6 years.

Fig. (25). *Euphausia superba.*

Luminous organs/Bioluminescence: It is commonly called as Antarctic krill or "light-shrimp " because it emits light. through its bioluminescent organs. In this species, one pair of luminous organs is seen at its eyestalk ; another pair are on the hips of the second and seventh thoracopods; and singular luminous organs on the four pleonsternites. These light organs have been reported to emit an yellow-green light up to 2–3 s periodically. Its light is comparable to that of a flashlight. A concave reflector present in the back of the luminous organ and a lens in the front seem to guide the light so produced by the animal. The whole light organ can be rotated by muscles which are also believed to direct the light to a specific area. The bioluminescent organs of this species contain several fluorescent substances and the major component has a maximum fluorescence at an excitation of 355 nm and emission of 510 nm [134].

Euphausia tenera

Image credit: Bárbara Santos Menezes (Reproduced with permission)

Depth of occurrence and Distribution: 0 – 1000 m; Cosmopolitan

Diagnosis & Size: In this luminescent species (Fig. **26**), eye is round and medium in size. Rostrum is short and its frontal plate is obliquely triangular. In the carapace, gastric dome is low, and is not angularly humped. It has a single lateral denticle . Adults reach a maximum length of 9 mm.

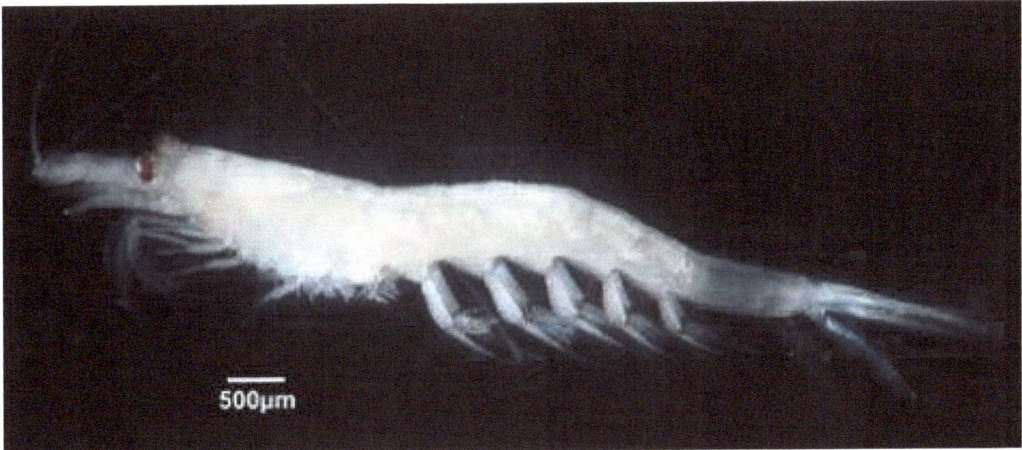

Fig. (26). *Euphausia tenera.*

Luminous organs/Bioluminescence: Not reported

Euphausia triacantha

Image credit: Jerzy Dzik & Krzysztof Jazdzewski

Depth of occurrence and Distribution: Depths down to 750 m; 250-500 m during day and 50-100 m at night; Polar, Antarctic ; Common in Southeast Pacific

Diagnosis & Size: It is commonly called as "spiny krill (Fig. **27**) . It has its characteristic medial spines on the abdominal tergites III-V. Distal segment of mandibular palp is short and wide. Adults attain a maximum length of 35 mm.

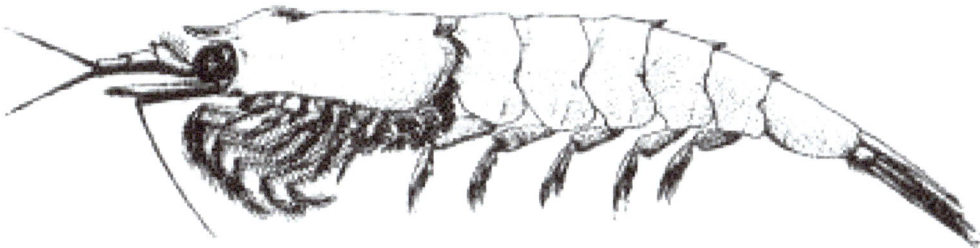

Fig. (27). *Euphausia triacantha.*

Luminous organs/Bioluminescence: Not reported

Meganyctiphanes norvegica

Image credit: Øystein Paulsen, Wikimedia

Depth of occurrence and Distribution: 100 and 400 m; restricted to the North Atlantic and subarctic Atlantic.

Diagnosis & Size: It (Fig. **28**) is commonly called as Northern krill. Eye is round and medium in size. Front margin of the carapace curves is slightly downward and is with a well developed post-ocular spine. A single pair of carapace denticles is seen near the edge. Adults reach a maximum length of 45 mm.

Fig. (**28**). *Meganyctiphanes norvegica..*

Luminous organs/Bioluminescence: On stimulation with short pulses, this species gave prolonged feeble glows which lasted 4-22 sec [61]. The fluorescent tetrapyrrole F of this species acts as a catalyst in the luminescent reaction. The terminal glows of this species showed a bimodal emission spectrum with a sharp primary peak at 476 nm and a second lower peak between 520 and 540 nm [132]. Further, the chemical stimulation with 5-hydroxytry ptamine yielded bioluminescent glow in this species and this glow was found to last a few minutes [135]. This translucent animal had low and spectrally flat reflectances averaging ~3% [130]. Its 5-hydroxytryptamine or related substance may be a part of the natural control system of the bioluminescence in this species [136].

Nematobrachion boöpis

Image not available

Depth of occurrence and Distribution: Deeper-living, below 400 m; North Atlantic; Irminger Sea

Diagnosis & Size: Upper lobe of eye is 1 to 2 times as wide as the lower lobe. There is no rostrum. In its carapace, frontal plate is broadly triangular, extending to the posterior limit of the eye. Carapace is without lateral denticles. It attains a maximum length of 25 mm.

Luminous organs/Bioluminescence: Some species of *Nematobrachion* have been reported to exhibit sexual dimorphism in the size or numbers of their ventral photophores [38].

Nematobrachion flexipes

Fisheries and Oceans Canada, Moira Galbraith, WoRMS

Depth of occurrence and Distribution: 100 - 600 m; Cosmopolitan

Diagnosis & Size: In this luminescent species (Fig. **29**) eye is large and the width of its upper lobe greater than the width of the lower. Rostrum extends to the anterior end of the eye as a slender keeled extension of the frontal plate. A small lateral denticle is seen near the lower margin of the carapace . Anterior part of the carapace has a well-developed keel.

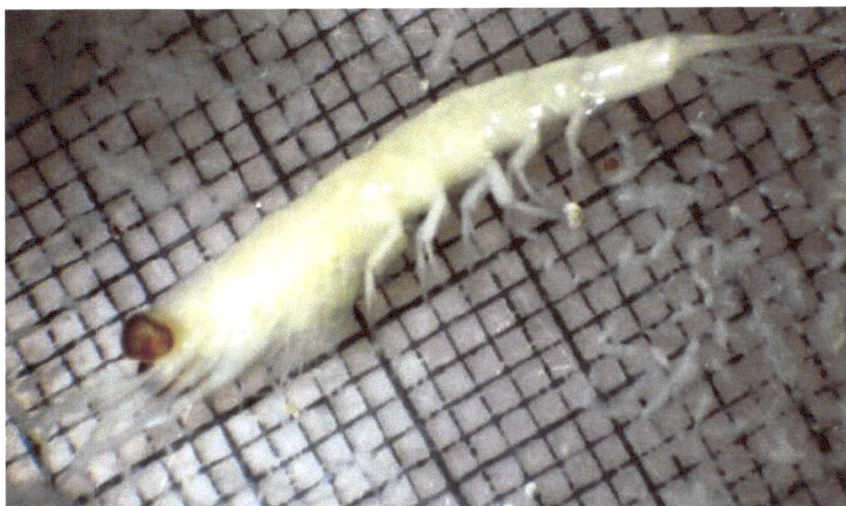

Fig. (29). *Nematobrachion flexipes.*

Luminous organs/Bioluminescence: In this species the males have been reported to possess 4 abdominal photophores and females two [137].

Nematobrachion sexspinosum

Image not available

Depth of occurrence and Distribution: 400 – 600 m; Tropical Eastern Pacific and Central Atlantic

Diagnosis & Size: Dorsal spines are seen on the posterior margins of 4th and 5th abdominal segments.It reaches a maximum length of 25 mm.

Luminous organs/Bioluminescence: Not reported

Nematoscelis atlantica

Image not available

Depth of occurrence and Distribution: 50 - 781 m; Tropical and Cosmopolitan

Diagnosis & Size: Eye of this luminescent species is bilobed and medium in size. Rostrum is long, slender and straight in both males and females sexes and are found extending to, or beyond the anterior end of the eye. A dorsal keel is low but distinct in its carapace. A lateral denticle is present at the carapace margin in both the sexes. It attains a maximum length of 12 mm only.

Luminous organs/Bioluminescence: Not reported

Nematoscelis difficilis

Image not available

Depth of occurrence and Distribution: Adults are commonly found in the upper 200 m. It is endemic to the North Pacific Ocean and Gulf of California

Diagnosis & Size: Eyes of this species are bilobed and their upper lobes are nearly the same width as the lower lobes. Its second thoracic leg is elongated. It attains a maximum length of 25 mm.

Luminous organs/Bioluminescence: Not reported

Nematoscelis gracilis

Image not available

Depth of occurrence and Distribution: 0 - 600 m; Tropical Indo-Pacific

Diagnosis & Size: Eye of this luminescent species is bilobed and medium in size.

Rostrum is short, acute in both males and females; and is somewhat upturned, extending forward to the midpoint of the obliquely tilted eye. There is a small, not well-defined dorsal keel in the carapace.No lateral denticles are seen in the carapace of adults. It attains a maximum length of 12 mm.

Luminous organs/Bioluminescence: Not reported

Nematoscelis megalops

Image credit: Bárbara Santos Menezes (Reproduced with permission)

Depth of occurrence and Distribution: 0 - 700 m; Subtropical Indo-Pacific, Atlantic

Diagnosis & Size: Eye of this species (Fig. **30**) is bilobed and large. Rostrum is short or absent in males. In females, it is usually long, slender, and downward curving. No lateral denticles are seen on the margin of the carapace in adults, but a pair of denticles is present in immature specimens. There are no dorsal spines or keels on its abdomen. Adults attain a maximum length of 26 mm.

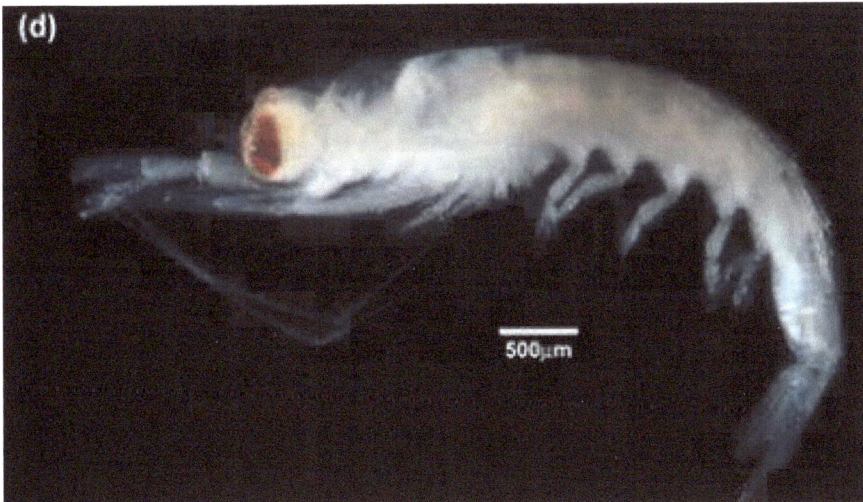

Fig. (30). *Nematoscelis megalops.*

Luminous organs/Bioluminescence: This translucent animal was found to have low and spectrally flat reflectances averaging ~3% [130].

Nematoscelis microps

Image not available

Depth of occurrence and Distribution: 0 - 820 m; Tropical and cosmopolitan

Diagnosis & Size: Eye of this species is bilobed and medium in size. Rostrum is short or absent in male; and acute in female and is reaching to the anterior limit of the eye. There is a low but prominent dorsal keel on the anterior part of the carapace. A denticle is seen on the lateral margin of the carapace in male, but it is absent in female. Adults may reach a length of 20 mm.

Luminous organs/Bioluminescence: Not reported

Nematoscelis tenella

Image not available

Depth of occurrence and Distribution: 0 - 600 m; Tropical and cosmopolitan species

Diagnosis & Size: Lower part of eye is smaller than upper part. Adults may attain a length of 21 mm.

Luminous organs/Bioluminescence: Some species of Nematoscelis have been reported to exhibit sexual dimorphism in the size or numbers of their ventral photophores [38].

Nyctiphanes australis

Image credit: William J. Dakin.Wikimedia

Depth of occurrence and Distribution: Common in coastal waters of south-east Australia

Diagnosis & Size: The name of this species *i.e* Nyctiphanes australis which means "southern winking light" is given to this species due to its luminescent characteristics. No other information is available.

Luminous organs/Bioluminescence: In this species (Fig. **31**), the bioluminescence is produced by a series of photophore organs present along the length of the body [138].

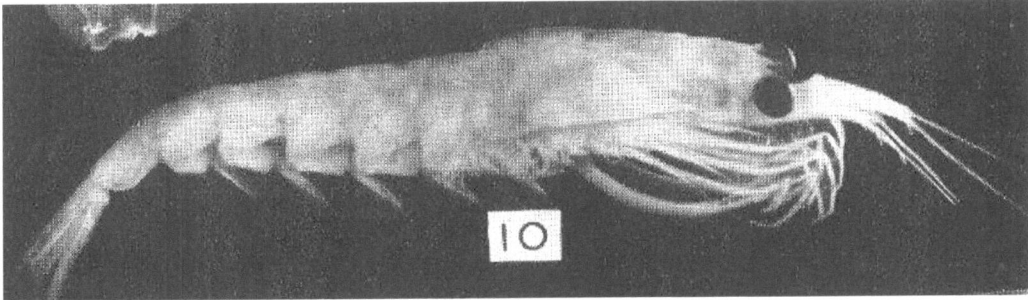

Fig. (31). *Nyctiphanes australis.*

Nyctiphanes capensis

Image not available

Depth of occurrence and Distribution: Coastal species (to about 250 m depth) dominant in the Benguela Current zooplankton community.

Diagnosis & Size: Eye of this species is rounded and larger in size. Rostrum is absent. Lateral carapace denticles are also absent.

Luminous organs/Bioluminescence: Not reported

Nyctiphanes couchii

Image not available

Depth of occurrence and Distribution: It is a neritic species common in the coasts of western Europe

Diagnosis & Size: Eye of this luminescent species is rounded and, larger in size. Rostrum is absent. No lateral carapace denticles in the carapace. Sixth segment of its abdomen has a small mid-dorsal posterior spine. It attains a length of 17 mm.

Luminous organs/Bioluminescence: Not reported

Nyctiphanes simplex

Image not available

Depth of occurrence and Distribution: It occurs to about 250 m depth, in shelf, slope and coastal waters in west of Baja California, Gulf of California and off southern California.

Diagnosis & Size: Eyes of this species are large and rounded. Rostrum and dorsal abdominal spines are absent. It has a leaflet on the lappet of its first antenna. It attains a maximum length of 16 mm.

Luminous organs/Bioluminescence: Not reported

Stylocheiron abbreviatum

Image credit: Bárbara Santos Menezes (Reproduced with permission)

Depth of occurrence and Distribution: 100 - 700 m; Tropical and cosmopolitan species.

Diagnosis & Size: Eye of this species (Fig. **32**) is pear-shaped and its upper lobe is much narrower than lower lobe. Rostrum is a horizontally produced plate which is reaching almost to the anterior limit of the eyes Carapace has a strongly domed gastric region and is with a small median keel. It attains a maximum length of 15 mm.

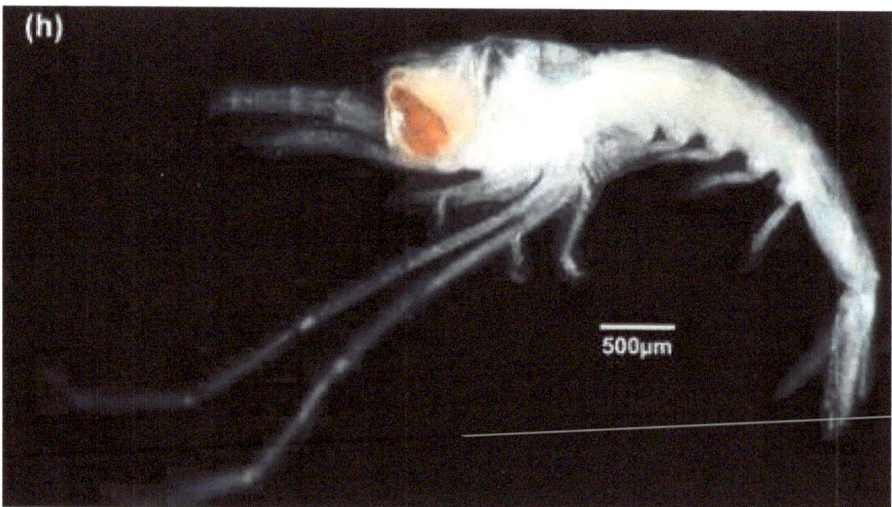

Fig. (32). *Stylocheiron abbreviatum.*

Luminous organs/Bioluminescence: In all the luminescent species of Stylocheiron the photophores which are responsible for emitting light are present on the eyes, the seventh thoracic limbs and the first abdominal segment [139].

Stylocheiron elongatum

Image not available

Depth of occurrence and Distribution: 181 - 820 m; Tropical and cosmopolitan species.

Diagnosis & Size: Eye of this luminescent species is cylindrical in appearance with upper and lower lobes which are almost equal in width. Rostrum is extremely short and acute in bothmales and females. Frontal plate is produced as an acute triangle which is narrowing to a short spiniform process at the apex. Carapace is with an elevated gastric area as a low dome and is with a low median keel. There are no lateral denticles. Adults ttain a maximum length of 13 mm.

Luminous organs/Bioluminescence: Not reported

Stylocheiron longicorne

Image not available

Depth of occurrence and Distribution: 0- 1600 m; Tropical and cosmopolitan species

Diagnosis & Size: Eye of this species is bilobed and its lower lobe is slightly wider than upper lobe. Its frontal plate is produced into a short acute rostrum which extends about to the anterior limit of the eye. Carapace is with domed gastric region and it has an indistinct short, low, median keel. There are no lateral denticles. It attains a maximum length of 10 mm.

Luminous organs/Bioluminescence: Not reported

Stylocheiron maximum

Image not available

Depth of occurrence and Distribution: 200- 2000 m; Tropical and cosmopolitan species

Maximum size & Diagnosis: Eye of this species is bilobed and its the upper lobe is almost equal in width to the lower lobe. Frontal plate is produced into an acute rostrum which almost reaches the anterior limit of the eyes. Carapace is with a domed gastric region and is with a small median keel There are no lateral denticles. Adults reach a maximum length of 25 mm.

Luminous organs/Bioluminescence: Not reported

Stylocheiron suhmii

Image not available

Depth of occurrence and Distribution: It is most common above 140 meters. Atlantic, Indian and Pacific oceans.

Diagnosis & Size: Eye of this luminescent species is narrow and its upper lobe bears 3 enlarged crystalline cones in a transverse row. Rostrum is slender and acute in females, and is extending to the anterior limit of the eyes. In males frontal plate terminates in a short, triangular plate. Gastric region of carapace is domed with a small-dorsal nub and rarely it is keel-like. It attains a maximum length of 7 mm.

Luminous organs/Bioluminescence: Not reported

Tessarabrachion oculatum

Image not available

Depth of occurrence and Distribution: This tropical pacific species occurs below 100m.

Diagnosis & Size: Eye of this luminescent species is very large and is with a constriction dividing both upper and lower lobes. Rostrum is absent. Its frontal plate is a short oblique triangle in shape. There is a low keel. In carapace, there are no denticles on the lateral margin. Adults may reach a maximum length of 20 mm.

Luminous organs/Bioluminescence: Not reported

Thysanoëssa furcilia

Image not available

Depth of occurrence and Distribution: Not reported

Diagnosis & Size: Its well developed rostrum is long to very long. Eyes are higher than broad and are constricted transversely.

Luminous organs/Bioluminescence: In this species the quantity of photons per flash was found to be in the order of 5x 10^9 [119].

Thysanoëssa gregaria

Image credit: Jerzy Dzik & Krzysztof Jazdzewski

Depth of occurrence and Distribution: 0-150 m; North and South Pacific, South Atlantic, and Indian ocean basins

Diagnosis & Size: Eye of this luminescent species (Fig. **33**) is bilobed with a transverse constriction. Width of its upper lobe is half the width of the lower lobe.

Rostrum is acute and is extending beyond the midpoint of the eye. There is a well-developed lateral denticle in its carapace. Adults may reach a maximum length of 17 mm.

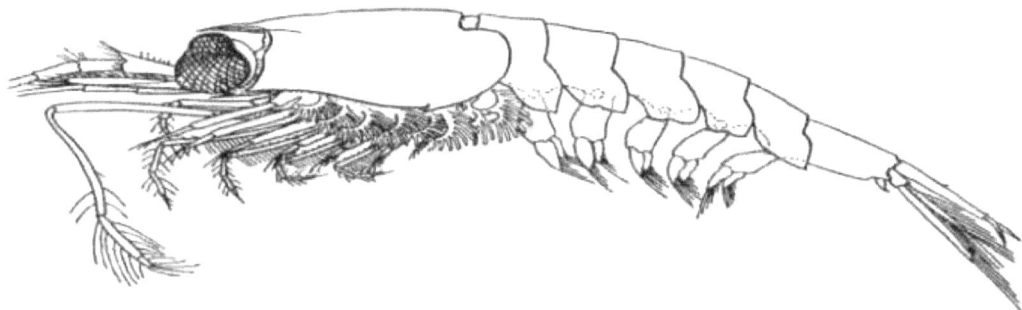

Fig. (33). *Thysanoëssa gregaria.*

Luminous organs/Bioluminescence: Not reported

Thysanoëssa inermis

Image credit: Jerzy Dzik & Krzysztof Jazdzewski

Depth of occurrence and Distribution: 0-300 m; North Pacific; arctic and subarctic waters of the Atlantic Ocean

Diagnosis & Size: Rostrum of this species (Fig. **34**) is narrow and acute, reaching beyond the eyes. Carapace has no lateral denticles. There is no dorsal keel on its abdominal segments. Eyes are almost circular and they may be slightly higher than broad.

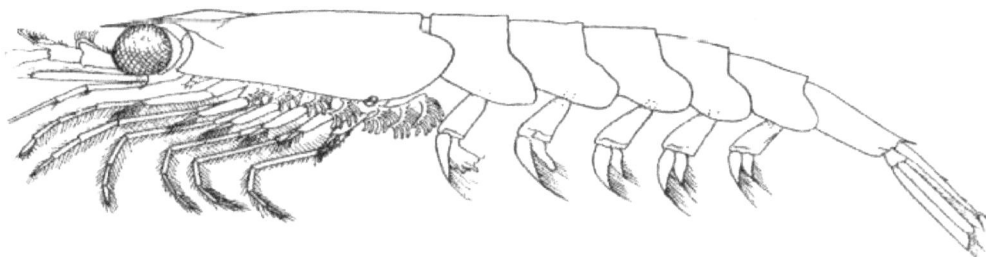

Fig. (34). *Thysanoëssa inermis.*

Luminous organs/Bioluminescence: Not reported

Thysanoëssa longicaudata

Image not available

Depth of occurrence and Distribution: 0-400 m; Arctic to Gulf of Maine

Diagnosis & Size: Eye of this luminescent species is bilobed with a transverse constriction. Rostrum is acute and is extending beyond the middle of the first segment of the of the antennula. Carapace is without denticles on the lateral margin. Adults may attain a maximum length of 16 mm.

Luminous organs/Bioluminescence: Not reported

Thysanoëssa macrura

Image credit: Jerzy Dzik & Krzysztof Jazdzewski

Depth of occurrence and Distribution: 0-400 m; circumpolar in the Antarctic

Diagnosis & Size: Eye of this luminescent species (Fig. **35**) is bilobed with a transverse constriction. Rostrum may be shorter in male than female and it is an acute triangle. There is a well-developed denticle on the lateral margin of the carapace. It may attain a maximum length of 30 mm.

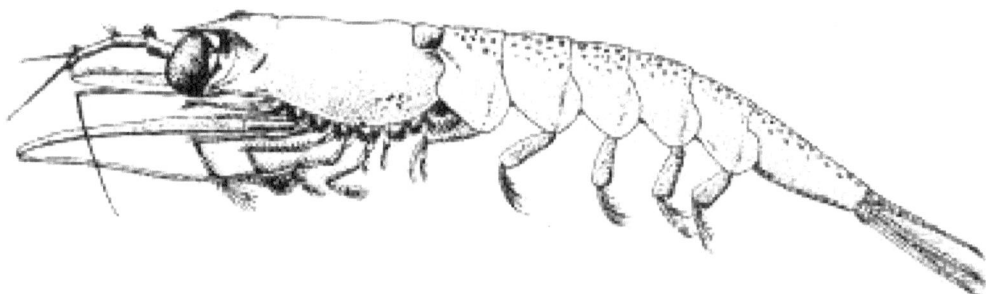

Fig. (35). *Thysanoëssa macrura.*

Luminous organs/Bioluminescence: Not reported

Thysanoëssa raschii

Image credit: Jerzy Dzik & Krzysztof Jazdzewski

Depth of occurrence and Distribution: Below 80 m; subarctic and Arctic seas

Diagnosis & Size: Rostrum of this species (Fig. **36**) is well developed and it forms a long plate which is broader in male than in female. Carapace carries a small lateral denticle. There are no keels on the abdominal segments. Eyes are sub-ovoid to almost spherical, without any constriction. It may attain a maximum length of 25 mm.

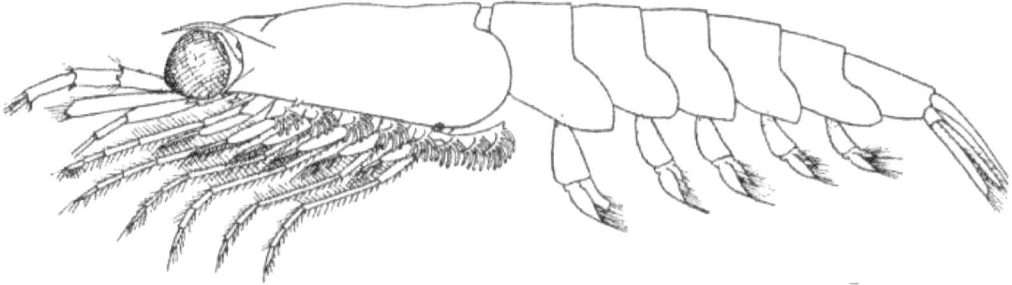

Fig. (36). *Thysanoëssa raschii.*

Luminous organs/Bioluminescence: The homogenate of this species has been reported to show a bimodal emission spectrum with a sharp primary peak at 476 nm and a second lower peak between 520 and 540 nm [132]. Its 5-hydroxytryptamine or related substance may be a part of the natural control system of the bioluminescence in this species [136].

Thysanopoda acutifrons

Image credit: Jerzy Dzik & Krzysztof Jazdzewski

Depth of occurrence and Distribution: Below 1000 m; Pacific and southwest Atlantic Oceans

Diagnosis & Size: Eye of this luminescent species (Fig. **37**) is small and spherical . Rostrum is a forward- and slightly upward-pointing process (tooth). Frontal plate is triangular and it extends to or beyond the eye usually. Carapace is without denticles on the lateral margins but a pair may be present in immature specimens.Eighth thoracic leg is extremely minute. It may attain a maximum length of 50mm.

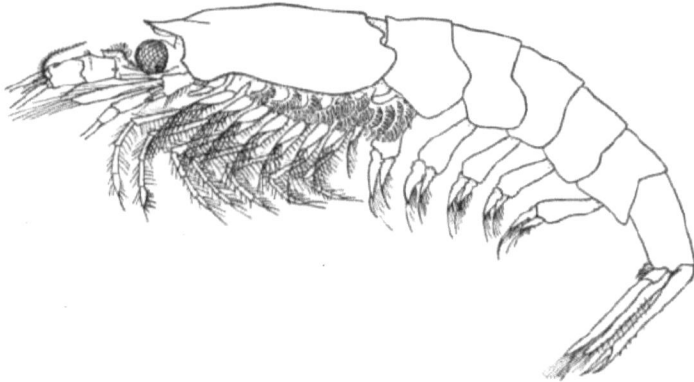

Fig. (37). *Thysanopoda acutifrons.*

Luminous organs/Bioluminescence: Not reported

Thysanopoda monacantha

Image credit: Jerzy Dzik & Krzysztof Jazdzewski

Depth of occurrence and Distribution: 100 - 1000 m; It is a tropical, cosmopolitan species.

Diagnosis & Size: Eye of this luminescent species (Fig. **38**) is medium sized. Rostrum is an acute, forward directed, triangular plate which is reaching almost to the anterior limit of the eye. Carapace: A longitudinal furrow is extending the length of carapace just above its lateral margin. Margin of the carapace bears a postero-lateral denticle. It may attain a maximum length of 32 mm.

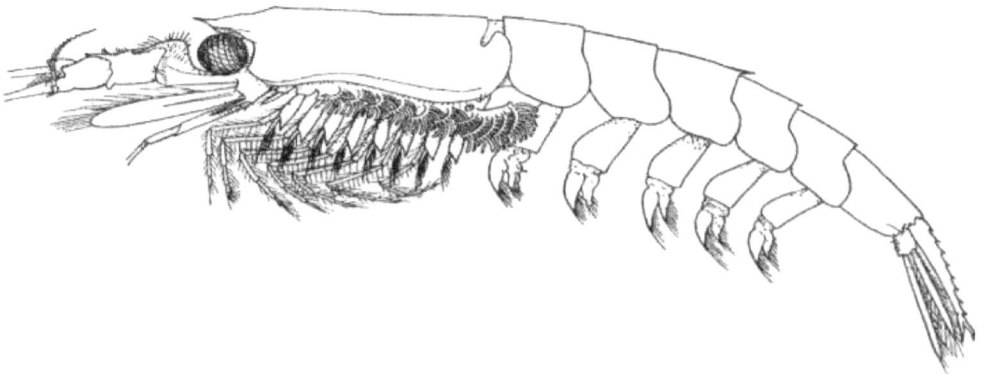

Fig. (38). *Thysanopoda monacantha.*

Luminous organs/Bioluminescence: Not reported

Thysanopoda microphthalma

Image not available

Depth of occurrence and Distribution: Below 500 m; Subtropical North Atlantic species.

Diagnosis & Size: Eye of this luminescent species is small. Rostrum is a short forward- and upward-pointing tooth.Carapace margin is without denticles but a single pair is present in immature specimens. Posterior margins of the 4th and 5th segments of abdomen are slightly pointed mid-dorsally. It may attain a maximum length of 41 mm.

Luminous organs/Bioluminescence: Not reported

Thysanopoda spinicaudata

Image credit: Jerzy Dzik & Krzysztof Jazdzewski

Depth of occurrence and Distribution: 2000-3000 m; Tropical Pacific and Indian ceans

Diagnosis & Size: Eye of this species (Fig. **39**) is very small and oval. Rostrum is a dorsal anterior end of the frontal plate extending to the most forward part of the eye and is provided with a long, strong vertical spine. Anterior margins of the frontal plate (lateral to the spine) are slightly upturned. Carapace is without lateral denticles. A cervical groove is found crossing the dorsal part of the carapace. Lateral furrows are continuous with this cervical groove. A low middorsal keel is seen on the carapace. It may attain a maximum length of 150 mm.

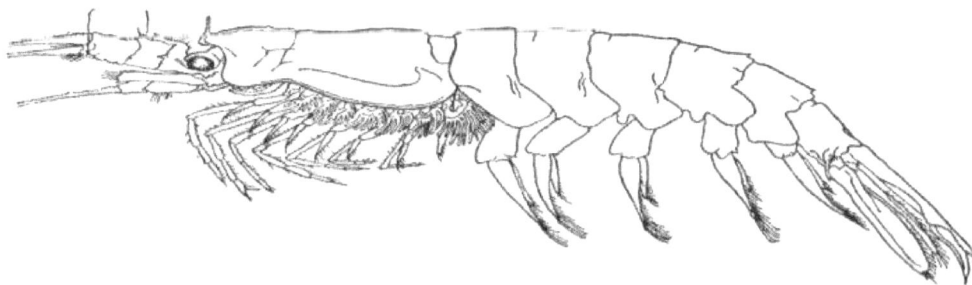

Fig. (39). *Thysanopoda spinicaudata.*

Luminous organs/Bioluminescence: In this species, the photophores present on the eyes only emiited light [139].

Thysanopoda tricuspidata

Image credit: Jerzy Dzik & Krzysztof Jazdzewski

Depth of occurrence and Distribution: 0 - 1000 m; Tropical Indo-Pacific and the Atlantic species.

Diagnosis & Size: Eye of this luminescent species (Fig. **40**) is medium in size. Rostrum is long and acute and is extending to the anterior limit of the eye. Carapace is a long, straight (forward-directed) spine which is extending up to the posterior limit of eye and is largely resembling a 2nd rostrum, above the frontal plate. Lateral margin of the carapace has two widely-separated denticles. The 3rd-6th segments of the abdomen bear mid-dorsal posterior spines. It may attain a maximum length of 25 mm.

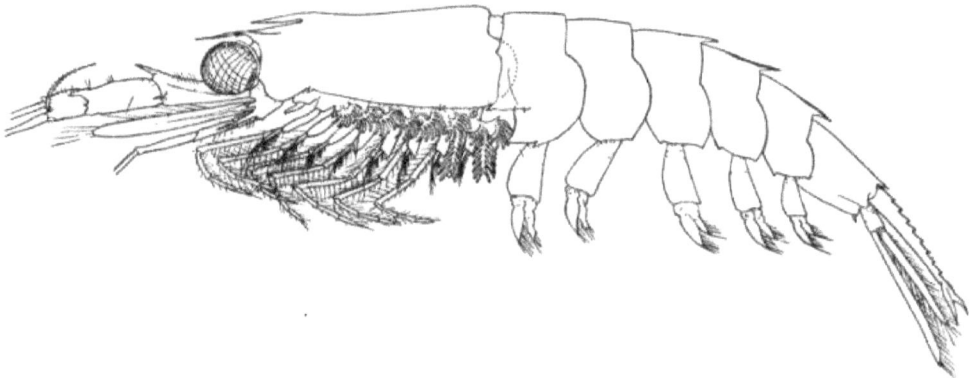

Fig. (40). *Thysanopoda tricuspidata.*

Luminous organs/Bioluminescence: Not reported

BIOLUMINESCENT PLANKTONIC MARINE DECAPODS

The name Lucifer (in Latin for "light bearer") was given to this genus because of these prawns are bioluminescent. Two species of this genus *viz.* Lucifer typus and Lucifer orientalis have been recognized [140]. It is however, reported that one species of the family Luciferidae *viz.* Lucifer typus was luminescent [13].

Luminescent lucifers

Class: Malacostraca; Order: Decapoda; Family: Luciferidae

Lucifer typus (= *Leucifer reynaudii*)

Image credit: Dr. Pierre NOËL (Reproduced with permission)

This luminescent species (Fig. **41**) has been reported to occur both in shallow and deep waters (0-700 m) of Atlantic, the Mediterranean and Indo-Pacific. Species of Luciferidae are the most peculiar shrimps in the world due to their aberrant appearance which includes a much compressed body, reduced appendages and branchia, and a curious copulatory organ. In this species, length of the neck is greater than the length of the eye stalk. Eyestalks are very slender; and eyes in male are larger than in female. Length of the rostrum reaches only to the base of the eye stalk. Outer spine of uropodal exopod is almost reaching distal end of exopod in both sexes. There are no gills. Females uniquely carry the fertilised eggs on her pleopods until they are ready to hatch. It has a common length of only 12 mm.

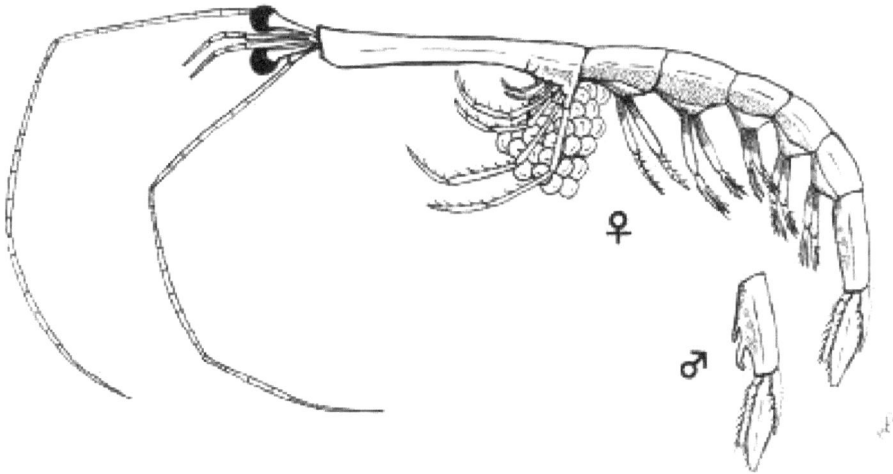

Fig.(41). *Lucifer typus.*

Bioluminescence: As this prawn is bioluminescent it is given the Latin name Lucifer *i.e.* "light bearer".

Lucifer orientalis

Image not available

This luminescent, oceanic prawn is distributed in the Indo-Pacific region of Eastern Africa to eastern Central Pacific. In this species eye stalks are very long, thin and cylindrical. Eye and eye stalks are slightly shorter than the distance between eye stalk base and labrum. Its rostrum extends slightly beyond the base of the eye stalks. Telson of male is short and rounded

Bioluminescence: Not reported

CONCLUSION

Though several families possess luminescent species, detailed studies have been made on the light organs and chemical reactions associated with light emission only in limited groups of planktonic crustaceans such as smaller copepods and larval stages of the euphausiids which have contributed to surface bioluminescence significantly.. Further analysis of bioluminescence of other crustacean groups in the world's oceans could be a clue for understanding the ecology and evolution of marine organisms.

Bioluminescent Marine Mollusc

Abstract: This chapter deals with the only luminescent species of planktonic marine nudibranch mollusc *viz. Phylliroe bucephala*,and its emission maxima and mechanism of bioluminescence.

Keywords: Emission maxima, Luminescent mollusc, *Phylliroe bucephala*.

INTRODUCTION

Molluscs are the most noteworthy group of marine organisms. They account for 7% of all the animals, while the phylum Mollusca is considered the second most important animal phylum with more than 52,000 identified and characterized species. The phylum Mollusca is gaining attention from the scientific community due to the number of their nutraceuticals of health importance. The promising potential of molluscs and products thereof the mitigation of human health problems call for more exploration and validation to fully utilize this enormous source of food drugs. Unlike freshwaters, where only a freshwater limpet (*Latia neritoides*) is bioluminescent, marine habitats are known for various luminescent molluscs such as nudibranchs, clams, cephalopods, and octopods. Amazingly, cephalopods possess the most advanced luminescence systems, and squids, particularly have been reported to have three or more types of luminous organs. Though several nektonic and benthic marine molluscs are luminescent, planktonic adult molluscs are rarely seen. This chapter, therefore, deals with the only species of planktonic nudibranch mollusc *viz. Phylliroe bucephala* (Fig. **1**).

Class: Gastropoda; Order: Nudibranchi; Family: Phylliroidae

Phylliroe bucephala

Image credit: Lydekker R., Wikimedia

This highly modified planktonic nudibranch is a deep water species found throughout the world and is common in the Atlantic, Pacific, and Mediterranean. The body of this luminescent species is laterally compressed, totally transparent, elongated, and fish or leaf-like. The tail is long and more than 16% of the body's length. It lacks papillae, but has one pair of well-developed tentacles (the rhinophores). The foot is reduced to a pedal gland and the anus is situated on the right lateral side in the centre of the body. It has 3 to 5 gonads and a parasitic life cycle. Its larval form attaches to the bell of the hydrozoan medusa *Zanclea costata* with its foot, where it feeds on the tissues of the bell. As the nudibranch reaches its maximum size of 550mm, this medusa shrinks in size, resulting in the adult nudibranch outgrowing the medusa after 10 days. This nudibranch then eats the host's tentacles and gets separated to live independently. Lastly, ithas a maximum length of 550 mm.

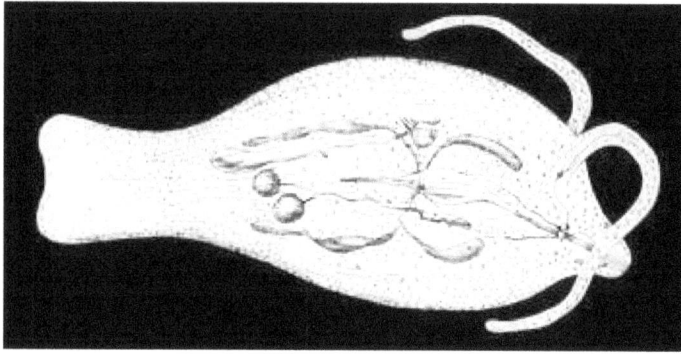

Fig. (1). *Phylliroe bucephala.*

Bioluminescence: It has bioluminescent markings on its body that are highly bioluminescent and emit flashes of light when disturbed, especially in the dark. In the luminescent molluscs, colenterizine (as luciferin) is present but there is no Ca2 – activated photoprotein [27].

CONCLUSION

Though several species of nektonic and benthic molluscs have been reported to be bioluminescent, reports on the bioluminescent species of planktonic marine molluscs are scanty except in a few species like the nudibranch, *Phylliroe bucephala*. This calls for intensive research on the diversity of planktonic, luminescent molluscs, their light organs, and light emission mechanisms.

CHAPTER 11

Bioluminescent Tunicates

Abstract: This chapter deals with the luminescent species of appendiculatians and thaliaceans; their emission maxima and mechanism of bioluminescence.

Keywords: Doliolid, Emission maxima, *Oikopleura*, *Pyrosoma*, Salp.

INTRODUCTION

The tunicates are small marine invertebrate animals, members of the subphylum, Tunicata of the Phylum, Chordata, found in great numbers throughout the seas of the world. Adult members are commonly embedded in a tough secreted tunic containing cellulose. While the bottom-dwelling and sessile (benthic) animals are less modified, the floating and swimming (planktonic and nektonic) animals are more advanced. While certain species of ascidians of Tunicata are consumed as food worldwide, several species of tunicates have been reported to contain a host of potentially useful chemical compounds such as didemnins, which are effective against various types of cancer, as antivirals, and as immunosuppressants. Bioluminescence takes different forms in the different groups of tunicates. Among the planktonic tunicates, Pyrosomes produce very brightly, long-lasting light by using bacteria to glow. On the other hand, most planktonic larvaceans embed luminous particles into their mucus "houses". Among the planktonic luminescent tunicates, the biology and ecology of certain species of *Folia* and *Oikopleura* (Oikopleuridae), *Pyrosoma* (Pyrosomatidae), *CycIosalpa, Helicosalpa* (Salpidae), *Doliolula* (Doliopsidae), *Pseudusa* (Doliolunidae) and *Paradoliopsis* (Paradoliopsida) are presented in this chapter.

Among these classes, Appendicularia and Thaliacea possess luminescent species of about 30 species (Table **1**). Special attention must be paid to identify the bioluminescent urochordates because they may trap luminous microorganisms, such as luminous dinoflagellates or luminous bacteria, which may induce extrinsic luminescence in these urochordates. For example, the brilliant blue-green luminescence of the colonial tunicate, *Pyrosoma* was earlier attributed to luminous bacteria, though, this symbiotic relationship is currently rejected. Its luciferin–luciferase reaction has not been demonstrated to date. Further, the bioluminescence of the appendicularians has been briefly reported currently., and It is considered as the coelenterazine-related luciferin–luciferase system of *Oikopleura labradoriensis* [4]. This calls for intensive research on the mechanism of bioluminescence in the planktonic urochordates.

Table 1. Bioluminescent tunicates [13]

Class	Order	Family	Species
Appendicularia	Copelata	Oikopleuridae	*Folia gracilis, Oikopleura albicans, O. cophocerca, O. dioica, O. drygalskii, O. gaussica, O. labradoriensis, O. mediterranea, O. parva, O. rufescens, O. valdiviae, O. weddelli, Stegosoma magnum*
Thaliacea	Pyrosomatida	Pyrosomatidae	*Pyrosoma atlanticum, P. spinosum, Pyrosomella verticillata*
-	Salpida	Salpidae	*Cyclosalpa affinis, C. floridana, C. pinnata, C.polae, C.bakeri, C. foxtoni, C. ihlei, C. quadriluminis f. paralleJa, C.q.f. quadriluminis, C. sewelli, HelicosaJpa komaii, H. virgula, H. younti,*
-	Doliolida	Doliopsidae	*Doliolula equus*
-		Doliolunidae	*Pseudusa bostigrinus*
-		Paradoliopsidae	*Paradoliopsis harbisoni*

Table 2. Emission maxima (λmax) of luminescent tunicates.

Species	Colour of light	(λmax)	Ref
Oikopleura dioica	Green	400 to 500 nm	[141]
O. labradorien	Green	400 to 500 nm	[141]
O. rufescens	Blue-green	---	[142]
Stegosoma magnum	Blue-green	---	[142]
Pyrosoma atlanticum	----	493nm	[10]
---	---	482 nm	[13]
P. a. giganteum	--	493 nm	[13]
P. spinosum	--	483-487nm	[10]
Pyrosoma spp.	---	490(540); 485-503nm	[10]
Pyrosomella verticillata	--	491, 483 nm	[13]
Doliolula equus	Blue	----	[143]

Emission Maxima in Luminescent Tunicates: The emission maxima (λmax) of the different species of luminescent tunicates have been reported to vary from 400 to 540 nm (Table **1**).

BIOLUMINESCENT APPENDICULARIANS (LARVACEANS)

Folia gracilis

Image not available

It is a circumglobal (except the Arctic) species. The body has an elongated trunk which is up to 0.5 mm long. Its buccal glands are very small, the endostyle is short, and the esophagus is long, entering the upper genital corner of the stomach. Moreover, the gonads are cup-shaped and parallel to the posterior wall of the genital cavity, and the tail is with a group of small spherical subchordal cells.

Bioluminescence: In this species, luminescence is assumed due to the presence of its photogenic oral glands [13].

Oikopleura albicans

Image not available

It is found distributed in the Indo-Pacific, Atlantic Ocean, and the Mediterranean. The trunk of this species is elongated and up to 5.0 mm long, the buccal glands are large, and the gonads are located adjoining the coil of the gut as a semicircular clasp protruding dorsally when ripe. The tail musculature is moderately broad with numerous, small, and stellar subchordal cells arranged in 2 lines over half of the tail length.

Bioluminescence: Certain oikopleurids possess paired oral glands that are considered the source of a "fluorescent" secretion. This secretion gets incorporated into the house and is responsible for the bioluminescence of the tunicate and its house. This species has been reported to give off spontaneous flashes of light while erecting a new house. Further, the finished house itself produces point sources of a fight when agitated [142]. This species also secretes long luminous filaments [13].

Oikopleura dioica

Image credit: Dr.David Fenwick (Reproduced with permission)

Bioluminescence: In luminescent larvaceans such as *Oikopleura dioica* (Fig. **1**), *Oikopleura labradoriensis, Oikopleura albicans, Oikopleura vanhoeffeni, Oikopleura rufescens,* and *Stegosoma magnum,* the granular inclusions which are probably present in their expanded house, accounted for the multiple, point-sources of light observed in flashing houses. In all these species, mechanical stimulation produced multiple, summated blue-green flashes from free individuals invested by house rudiments and from empty houses which were thoroughly free from exogenous luminescent bacteria or other luminescent microorganisms. Further in these species, the light is produced by clusters of 1 to 2 μm fluorescent granules that form intricate, species-specific patterns of inclusions in the house rudiment [141]. All the six known luminescent species of larvaceans possess fluorescent and luminescent house rudiment inclusions and oral glands [142].

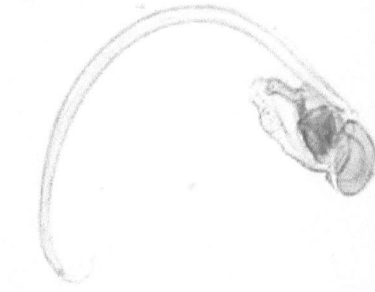

Fig. (1). *Oikopleura dioica.*

Oikopleura labradoriensis

Image not available

This medium-sized, luminescent appendicularian is found distributed in the Arctic, northern North Atlantic, and northern North Pacific Oceans. The trunk is elongated, the tail has a long series of sub-chordal cells near the tip, the left lobe of the stomach is trapezoidal, and the buccal glands are slightly large and oval in shape. Moreover, a single ovary and two paired testes are seen posteriorly in the trunk and tts body length can go up to 2.6 mm.

Bioluminescence: This species uses a coelenterazine+luciferase system to produce of light [8].

Oikopleura rufescens

Image not available

This species is found distributed in the Indo-Pacific, the Atlantic Ocean and the Mediterranean. The trunk is compact and can go up to 2.3 mm long. The buccal glands are spherical and fairly large, and the gonads are located adjoining the gut coil protruding dorsally, and tapering toward the posterior end. Lastly, the tail has a spindle-shaped subchordal cell, and its musculature is narrow.

Bioluminescence: It is endogenously luminescent and gives blue-green flashes upon stimulation of free animals with house rudiments [142].

Oikopleura (Vexillaria) vanhoeffeni (= Oikopleura vanhoeffeni)

Image not available

This cold water, luminescent species rarely occurs in the south of the Shetland Islands, UK. It is a fairly large luminescent species reaching a total length of 7 mm. The compact trunk, the round left lobe of the stomach, and thea membraneous circumoral lip is present. Moreover, both ovary and testes form a round structure behind the stomach and oesophagus. Further, the endostyle is long and narrow, the tail is long, with a strong chord and a medium-sized muscular band. Lastly, numerous small sub-chordal glands are seen on the right side of the proximal third part of the tail.

Stegosoma magnum

Image not available

This subtropical species is found distributed in the Indo-Pacific, the Atlantic Ocean, and the Mediterranean at a depth range of 0 – 1300 m.

Bioluminescence: It is endogenously luminescent and gives blue-green flashes upon stimulation of free animals with house rudiments [142].

BIOLUMINESCENT THALIACEANS

Order: Pyrosomida

Bioluminescent pyrosomes

Pyrosoma atlanticum

Image credit: Rhododendrites, Wikimedia

It (Fig. **2**) is a temperate species found distributed in all the world's oceans, usually at depths below 250 m.It is a cylindrical colony of about 60 cm length and is with a tough consistency . It is provided with numerous longer or shorter truncate test processes, tapering into an acanthose, backwards pointing tip. Open end of colony is with a tight diaphragm.In the colony, zooids are tightly packed. Zooids are rounded in shape and sometimes they are more angular or even triangular. Length of each zooid may be up to 8.5 mm. Sexually mature zooids are seen in colonies of more than 4-6 cm length. Colour of the colony is pink or yellowish pink.

Fig. (2). *Pyrosoma atlanticum* .

Bioluminescence: In life, the colonies of this species can glow due to their individual zooids which are believed to be bioluminescent. These colonial tunicates have been reported to generate brilliant, sustained bioluminescence in response to light or mechanical stimulation [144]. In their studies relating to *Pyrosoma atlanticum* and *Pyrosomella verticillata*, they also mentioned that each zooid within a colony detects light and emits bioluminescence in response. Further, photic stimulation of 1.5×10^9 photons$\{$middot$\}$s$^{-1}\{$middot$\}$cm^{-2}, at wavelengths between 350 and 600 nm, induced bioluminescence in these two species, with the maximum response induced at 475 nm. The photic-excitation half-response constant was found to be 1.1×10^7 photons$\{$middot$\}$s-1$\{$middot$\}$ cm^{-2} at 475 nm for *Pyrosoma atlanticum* and *Pyrosomella verticillata* on the other hand, had a significantly higher half-response constant of 9.3×10^7 photons$\{$middot$\}$s$^{-1}\{$middot$\}$cm^{-2}. However, individual zooids within a colony had different half-response constants. It was also found that the strength of stimulus influenced recruitment of zooids and, in turn, luminescent duration and quantum emission. Further, in these species, repetitive, regular mechanical or electrical stimulation produced rhythmic flashing which was characterized by alternating periods of high and low light intensities [144].

The gigantic blooms of pyrosomes which are also called as " fire bodies" play important ecological and biogeochemical roles in oceans [145]. The biochemistry of light production in pyrosomes is unknown, but it has been hypothesized to be bacterial in origin. These authors found that mixing coelenterazine—a eukaryote-specific luciferin—with *Pyrosoma atlanticum* homogenate produced light. And it was suggested that the putative luciferase (PyroLuc) of this species produced light by combining with the coelenterazine. The light produced by this species following mechanical stimulation is shown in Fig. (**3**).

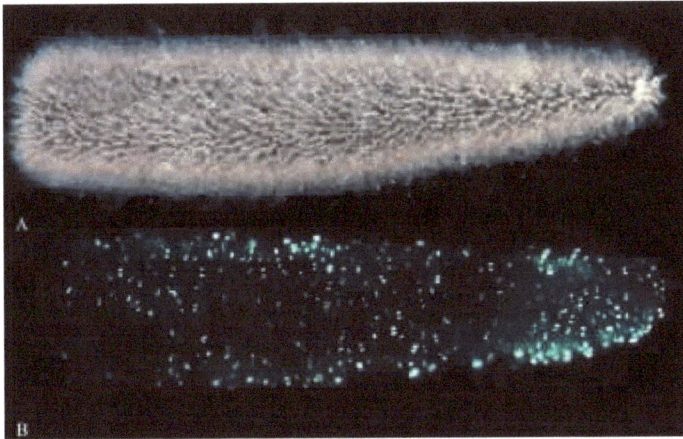

Fig. (3). *Pyrosoma atlanticum* producing bioluminescence.

Image credit: Scientific Reports CC

Pyrosoma spinosum

Image not available

It is often found drifting in the shallow waters off central and southern New South Wales; and central Arabian Sea . It is commonly called as "Giant Pyrosome" of "Unicorn of the Sea" which is a free-floating, colonial tunicate which is made of thousands of identical clones. It forms a hollow cylindrical structure of about 30 m long and it is wide enough for a person to enter. Each individual clone is a small, complete animal . As Tthese clones have a notochord ("spinal" chord), these animals are included under vertebrates.

Bioluminescence: These giant pyrosomes are bioluminescent and the light produced by these animals is bright and long lasting. These are also called as "fluorescent cheetos of the sea" especially when they are near the surface [146].

Pyrosomella verticillata

Image not available

This subtropical species is common in the Northwest Pacific particularly in Japan. Colony is finger-like and oval and is often flattened. It may reach a size up to 5 cm long and 3 cm in diameter. Colonial wall is transparent, soft and colorless. Zooids of the colony have very short oral siphon and short cloacal siphon.

Bioluminescence: This species has been reported to produce light by mechanical stimulation (Fig. **4**).

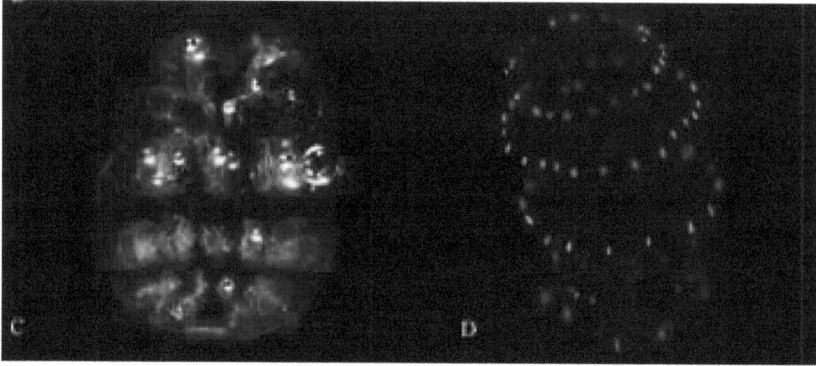

Fig. (4). Bioluminescence of *Pyrosomella verticillata.*

Image credit: Scientific Reports CC

Order: Salpidae

Bioluminescent salps

The species of salps are approximately 13 cm long and are barrel-shaped organism resembling streamlined jellyfish. The live in mid-ocean waters where they filter the seawater for their food . Some species of salps are bioluminescent and exude flashes of light [147].

Cyciosalpa affinis

Image not available

It is found distributed in the temperate and tropical waters and is common in California waters and occasionally in the Gulf of Alaska. While the size of aggregate generation is up to 46 mm, solitary generation is up to to 74 mm long. Aggregate generation zooid possesses a very flaccid tunic, and the intestine is like a broad loop. Solitary generation zooid also possesses a very flaccid tunic and elongate intestine or gut. Aggregate generation forms circular chains connected to one another.

Luminous organs: Bioluminescence has been reported in *Cyclosalpa* spec. (aff. *affinis*). The solitary zooid of this variety possesses one pair of luminous organs which are roundish and of a very light colouring. These luminous organs are positioned between M IV and M V [148].

Cyclosalpa bakeri

Image not available

This species has been reported to be present in the upper 150 m at night and is mostly deeper by day. It is found distributed in tropical and temperate waters, and as far north as the Gulf of Alaska. While the size of aggregate generation is up to 26 mm;, solitary generation is up to 47 mm. Aggregate generation has a characteristic intestinal loop, trailing testis, and caecum within posterior projections. Solitary generation is barrel-shaped. Aggregate generation forms circular chains connected to one another, and possesses posterior projections.

Luminous organs: Solitary generation of this species is with 3-5 light organs which are located on lateral sides running perpendicular to the horizontal body muscles. In the individuals where five strongly developed pairs of luminous organs are pr3sent, they lie between M I and M VI and one small pair between the intermediate muscle and M 1. These light organs are often broken or indented halfway between the body muscles, and this gives the appearance as if the specimen has 11 pairs of light organs (the first small pair is never broken) instead of 6. These light organs possess ring of luminous granules [13, 148].

Cyclosalpa floridana

Image not available

It is a tropical species found at depths of 0-1000 m in the Northern Atlantic and Western Central Pacific especially in Europe. Solitary zooids are up to 13 mm long. Body muscles are interrupted dorsally. MI to MII are fused dorsally. Intermediate muscle laterally joining MI. MI to MV are fused ventrally into a single mass. MVI is interrupted but MVII is continuous ventrally. Aggregate zooids are up to 9.5 mm long. Apparently there are only 3 body muscles (probably MI and MII are fused). Intermediate muscle is joining MI laterally; and MIII and MIV are fused dorsally. MI (MII) and MIII to MIV are converging but are not contiguous in the mid-dorsal line.

Luminous organs: In this species, luminous organs that are present only in solitary forms are weakly developed. They are more or less continuous, paired mass positioned between M I and M VI or between M II and M VI. Luminous organs are completely absent in aggregate forms [148].

Cyclosalpa foxtoni

Image not available

It is a circum-(sub)tropical species found distributed in Atlantic. Test of solitary zooids is very thick, globular, voluminous and is quite transparent. Size of these zooids may be up to 37 mm . Body muscles are arranged exactly like those of C. bakeri. However, the muscles of this species are significantly narrower than in C. bakeri. Dorsal tubercle is smooth, simple and G-shaped.

Luminous organs: In contrast to all known cyclosalpas, luminous organs are in the form of three to four pairs of weakly developed dots of luminous material which are situated on the body muscles M II - M V, instead of between them [148].

Cyclosajpa ihlei

Image not available

Test of the solitary zooids of this species is thin, flabby, and is closely adhering to the body. Size of these zooids may be up to 51 mm. Body muscles are interrupted dorsally and ventrally and there are no longitudinal muscles.

Luminous organs: There are 5 pairs of luminous organs which are in the form of strongly developed stripes between the muscles M I - M VI.I in addition to these, an extra pair of luminous organs is also present between M V and M VI parallel to the normal pair [148].

Cyclosalpa pinnata

Image not available

This subtropical species is found at depths of 0 – 90 m in Eastern Indian Ocean, Atlantic Ocean and the Mediterranean, especially in Europe. Solitary zooids of this species are up to 75 mm long. Body muscles are interrupted dorsally. Dorsal longitudinal muscles are absent. Five pairs of light organs between MI and MVI. Dorsal tubercle moderately convoluted. Aggregate zooids are up to 64 mm long . In these zooids, body muscles are symmetrically arranged ; MI and MII are fused dorsally; and MIII and MIV are converging but not contiguous dorsally.

Luminous organs: In the solitary zooids there are five pairs of strongly developed light organs (as luminous stripes) and are present between MI and MVI. In the aggregate zooids, one pair of light organs is present between the muscles MII and MIII [148].

Cyciosalpa polae

Image not available

Test of solitary zooids is soft and thin. Size of these zooids is up to 40 mm. In these zooids, body muscles are interrupted dorsally and ventrally with the exception of M VI. Fusion of M VI is complete.

Luminous organs: There are 5 pairs luminous organs in the solitary zooids of this species. They are strongly developed and are located between the muscles, M I - M VII [148].

CycIosalpa quadriluminis forma *quadriluminis*

Image not available

Only aggregates are known in this species. Test of these zooids is thick. Size of these zooids is up to 29 mm. Body muscles M I - M II and M III - M IV are fused dorsally.

Luminous organs: In the aggregate zooids of this species, there are 2 pairs of luminous organs which are lying between the muscles M II - M III and M III - M IV. The anterior pair is about two times as long as the posterior pair [148].

Cyclosalpa Quadriluminis forma *parallela*

Image not available

Only aggregate generations of this form are known. Test of these zooids is moderately thick, and transparent. Size of these zooids may be up to 37 mm.

Luminous organs: There are 5 pairs of strongly developed luminous organs and these organs are lying between the muscles M I and M VI [148].

Cyclosalpa sewelli

Image not available

This tropical species is found in Indo-Pacific areas such as Australia and Mexico. Test of the solitary zooids of this species is soft and thin. Size of these zooids is up to 24 mm. All body muscles are interrupted dorsally.

Luminous organs: The solitary zooids of this species possess 4 pairs of luminous organs which are located between the muscles M II and M VI [148].

Helicosalpa komaii

Image not available

It is a subtropical, oceanic species found in the Northwest and Northeast Pacific especially in Taiwan. Test of the solitary zooids of this species is flabby and is closely adhering to the body. Size of these zooids is up to 230 mm. Body muscles M I and M II get fused near the mid dorsal line. There is only one dorsal longitudinal muscle linking M I/M II with M VII. Two ventral longitudinal muscles are found linking the intermediate muscle with M VII. All body muscles are fused with the ventral longitudinal muscles.

Luminous organs: In the solitary zooids of this species, the luminous organs are in the form of a continuous line on each side of the body from the muscles M I to M VI [148].

Helicosalpa virgula (= *Cyclosalpa virgula*)

Image not available

This cosmopolitan species is found in the waters of tropical to subtropical regions. Test of solitary zooids of this species thin, voluminous. Size of these zooids is up to 180 mm. Muscle M I is linked with M VI by a paired dorsal longitudinal muscle; and M I and M II are fused near the mid dorsal line into the dorsal longitudinal muscles. Ventral longitudinal muscles (between M I and M V) are interrupted on the level of M III. Body muscles of these zooids are not fused with the ventral longitudinal muscles.

Luminous organs: In the solitary zooids of this species, luminous organs are in the form of a continuous line on each side of the body from M I to M VI. This line may also have a slight extension in front of M 1. In the aggregate zooids, the luminous organs are absent [148].

Helicosalpa younti

Image not available

This subtropical species is found in Southeast Pacific especially in Chile. Test of the solitary zooids of this species is enormous compared to its body and is globular and flabby. Size of these zooids is up to 142 mm. Body muscles are arranged as in H. virgula. Intermediate muscle is ventrally interrupted, whereas it is continous in H. virgula.

Luminous organs: In this species, the luminous organsare in the form of 6 pairs of luminous stripes (5 strongly developed and one small pair) and these stripes are positioned between the intermediate muscle and M VI [148].

Order: Doliolidae

Bioluminescent doliolids

Doliolula equus

Image not available

This Subtropical species is found at a depth range of 160 – 400 m in Eastern Central Pacific, especially in USA. It is an ovate form of doliolid and is with five girdling muscle bands. Its third band consists of two, L-shaped segments which are not connected to each other dorsally or ventrally. All other muscle bands are complete rings. As in the other species of the suborder Doliopsidina, only blastozooids are known in this species. Individual zooids are seen along the stolon (lace) and gonozooids are hermaphroditic. This species is bioluminescent and about one zooid in ten is peppered with orange pigment spots.

Bioluminescence: This species has been reported to produce a diffuse, blue glow upon mechanical stimulation. The light so produced outlined the body and originated in the tunic rather than its interior body. In several cases, the light pulses were of short duration, and were lasting only a few seconds after the stimulus. However, with continuous stimulation, the glow persisted and the intensity of the output could be related to the level of the stimulus. Rarely, a sustained glow from a single individual in a colony was found to last for more than 30 seconds after stimulation [143].

Paradoliopsis harbisoni

Image not available

This luminescent species has been reported from Western Atlantic *viz*. Bahamas [12]. It has a maximum length of 9 mm. Its wide-open buccal and atrial siphons are visible. Body length is slightly longer than high; atrial siphon is long; buccal siphon is wide; and buccal vestibule is capacious. It has conspicuous red-orange and yellow-gold pigmentation in the U-shaped digestive canals.

Pseudusa bostigrinus

Image not available

This luminescent species has been reported from the Eastern Central Pacific [12]. No other information is available for this species.

CONCLUSION

Though considerable reports are available on the diversity of luminescent fauna of the subphylum Tunicata, the biochemical mechanism of bioluminescence of a few such as that of Pyrosoma has been studied. This calls for further research on similar studies with other luminescent genera in order to understand the importance of this planktonic group in the marine food chain.

CHAPTER 12

Marine Bioluminescence and Biotechnology

Abstract: The role of bioluminescent marine plankton in the fisheries and the health of the oceans is fairly well known. Recent research has shown that this light could be of great use in therapeutic and biotechnological applications. While the marine bioluminescent, planktonic crustaceans are helpful in the treatment of cancers, other groups of marine plankton have their potential biotechnological applications, including hygiene control and mapping out pollution in ecosystems. However, further intensive research is needed on this vital aspect, especially when identifying lesser-known bioluminescent planktonic groups and their biomedical and biotechnological applications for the benefit of human society.

Keywords: ATP sensing, Bioluminescent imaging, CAR T-cell therapy, Hygiene control, Marine pollution.

INTRODUCTION

The phenomenon of bioluminescence has been reported to be helpful in biomedical applications, especially wheninvestigating and monitoring cell health and the impact of drug treatments. For instance, if one compares the light given off by cells in treated wells against untreated wells (light is a proxy for the number of live cells), the amount of light given off by the untreated samples would be higher than the treated wells if the chemotherapeutic is effective. The bioluminescent proteins, including the luciferases may serve as biosensors in the targets of drugs used to treat a wide array of disorders and diseases, including diabetes, allergies, pain, and hypertension [149]. Further, these biosensors could have a major impact on new drug development. Additionally, bioluminescence can also be used within biological systems for monitoring water systems to ensure high-quality drinking water, for detecting proteins, for determining whether a patient's blood sample contains antibodies to COVID-19, and for vaccine research and development. These findings may lead up to the developments in synthetic luciferins and engineered luciferases for other biotechnological applications, including ATP sensing, hygiene control in the fish and milk industries, mapping out pollution in ecosystems *etc*. It can also be reasonably expected that developments in these areas will feed into the research and applications of bioluminescence in biotechnology.

THE ROLE OF MARINE BIOLUMINESCENT PLANKTONIC CRUSTACEANS IN CANCER IMMUNOTHERAPIES

The marine bioluminescent planktonic crustaceans have been reported to play a major role in the treatment of cancers. The enzyme, luciferase from these organisms, when introduced into cancer cells relating to chronic myelogenous leukemia, acute myelogenous leukemia, and Burkitt lymphoma, was found to leak out as the cells die, leaving a visible glow. This assay has been reported to accurately recognize the death of a single cancer cell in 30 minutes [149]. It is also reported that this technique can also be employed in CAR T-cell therapy (chimeric antigen receptor T-cell therapy) in which a patient's T-cells (a type of immune system cell) taken from his blood are changed in the laboratory so they can attack cancer cells. In the development of CAR T cells, the proteinof bioluminescent crustaceans' luciferase can play a leading role in the future (chimeric antigen receptor T-cell therapy [150].

Bioluminescent Imaging

The green fluorescent protein (GFP), first isolated from jellyfish and subsequently from corals, sea anemones, zoanthids, copepods, and lancelets, exhibits bright green fluorescence when exposed to light in the blue to ultraviolet range. This GFP has a wide application in bioluminescent Imaging (BLI) to study the interaction of infectious bacteria and fungi with living cells [151].

MARINE BIOLUMINESCENT PLANKTON AND THEIR BIOTECHNOLOGICAL APPLICATIONS

Though a total of about 30 bioluminescent systems have been identified, the luciferin-luciferase pairs of only 11 systems have been characterised to date. As the different luciferin-luciferase pairs have different light emission wavelengths they are deemed suitable for various applications [152]. Bioluminescent organisms may also serve as a target for many areas of research. Presently, luciferase systems are widely used in the field of genetic engineering. Certain applications of engineered bioluminescence include: i) glowing trees to light up highways and save electricity, ii) crops and domestic plants which luminesce when they need watering, and iii) novelty pets that luminesce (rabbits, mice, fish *etc.*) [153].

Hygiene Control

The ATP-bioluminescence assay based on the firefly bioluminescence system is presently used in applications relating to hygienic control. This technology is applied to monitor the cleanliness of surfaces in healthcare facilities such as hospitals and clinics; and food and dairy industries.Similar applications would also be made possible in the future with bioluminescent marine plankton if and when adequate research is attempted [152].

MAPPING OUT POLLUTION IN ECOSYSTEMS

The common application of bioluminescence in ecotoxicology and pollutant monitoring is ATP quantification using the firefly bioluminescence ATP assay in the marine environment. Similar applications would also be made possible in the future with bioluminescent marine plankton [152].

FUTURE ISSUES IN MARINE BIOLUMINESCENCE

i): Though the genes for many luciferases derived from marine-luminescent organisms are known, the mechanisms of luciferin biosynthesis are yet to be known. This will be a promising area for future research in marine bioluminescence.

ii): The chemistry of luminescence for many organisms such as planktonic polychaetes and tunicates, is still completely unknown. Intensive research is therefore needed on this aspect.

iii): In order to thoroughly understand the marine ecosystem dynamics, harmful algal blooms, and how and why plankton populations fluctuate over time, improvements in remote and automated methods of detecting oceanographic-scale bioluminescence (satellites and bathyphotometers) are the need of the hour.

CONCLUSION

It is worth mentioning here that the function of bioluminescence has been investigated much less than any other topic in the field. Though bioluminescence has been found to be very useful in various fields, including medicine, biology, physics, and engineering, intensive research is greatly needed on aspects such as the origins and mechanisms of bioluminescence, luciferin and luciferase systems, in less studied luminescent marine organisms, to derive advanced applications of bioluminescence in biotechnology across various fields and make them accessible to particular target groups.

REFERENCES

[1]   Hewitt L, Lacey E. Luciferin: The glowing group of molecules responsible for bioluminescence. 2019. http://www.chm.bris.ac.uk/motm/luciferin/luciferinh.htm

[2]   Ohmiya Y, Kojima S, Nakamura M, Niwa H. Bioluminescence in the Limpet-Like Snail, *Latia neritoides*. Bull Chem Soc Jpn 2005; 78(7): 1197-205.
[http://dx.doi.org/10.1246/bcsj.78.1197]

[3]   Bioluminescence. https://www.nationalgeographic.org/encyclopedia/bioluminescence/

[4]   Oba Y, Stevani CV, Oliveira AG, Tsarkova AS, Chepurnykh TV, Yampolsky IV. Selected Least Studied but not Forgotten Bioluminescent Systems. Photochem Photobiol 2017; 93(2): 405-15.
[http://dx.doi.org/10.1111/php.12704] [PMID: 28039876]

[5]   The Black Dragonfish. https://biolum.eemb.ucsb.edu/organism/dragon.html

[6]   Johnsen S. Bioluminescence. https://oceanexplorer.noaa.gov/explorations/15biolum/background/biolum/biolum.html

[7]   Staples RF. The distribution and characteristics of surface bioluminescence in the oceans. US Naval Oceanographic Office Washington DC Technical report 184 1966. https://apps.dtic.mil/dtic/tr/fulltext/u2/630903.pdf
[http://dx.doi.org/10.21236/AD0630903]

[8]   Haddock SHD, Mastroianni N, Christianson LM. A photoactivatable green-fluorescent protein from the phylum Ctenophora. Proc Biol Sci 2010; 277(1685): 1155-60.
[http://dx.doi.org/ 10.1098/rspb.2009.1774] [PMID: 20018790]

[9]   Miller SD, Haddock SHD, Elvidge CD, Lee TF. Detection of a bioluminescent milky sea from space. Proc Natl Acad Sci USA 2005; 102(40): 14181-4.
[http://dx.doi.org/10.1073/pnas.0507253102] [PMID: 16186481]

[10]   Herring PJ. The spectral characteristics of luminous marine organisms. Proc R Soc Lond B Biol Sci 1983; 220(1219): 183-217.
[http://dx.doi.org/10.1098/rspb.1983.0095]

[11]   Kahlke T, Umbers KDL. Bioluminescence. Curr Biol 2016; 26(8): R313-4.
[http://dx.doi.org/10.1016/j.cub.2016.01.007] [PMID: 27115683]

[12]   Martini S, Haddock SHD. Quantification of bioluminescence from the surface to the deep sea demonstrates its predominance as an ecological trait. Sci Rep 2017; 7(1): 45750.
[http://dx.doi.org/10.1038/srep45750] [PMID: 28374789]

[13]   Poupin J, Cussatlegras A, Geistdoerfer P. Bioluminescent Marine Plankton: Documented inventory of species and assessment of the most common forms of the Iroise Sea. Naval School Oceanography Laboratory Research report 1999.

[14]   Santhanam R. Marine Dinoflagellates. Nova Science Publishers. USA. 2015.

[15]   Bosch-Belmar M, Milisenda G, Basso L, Doyle TK, Leone A, Piraino S. Jellyfish Impacts on Marine Aquaculture and Fisheries. Rev Fish Sci Aquacult 2021; 29(2): 242-59.
[http://dx.doi.org/10.1080/23308249.2020.1806201]

[16]   Rajkumar M, Aravind R, Praveen Raj J, Artheeswaran N, Pandey AP. Commensalism between jellyfish and juveniles of carangids in coral reef habitats of Palk Bay, India. Eco Env Cons 2014; 20(1): 139-41.

[17]   Baker S. The Secrets of Jellyfish. 2015. https://www.bcmag.ca/the-secrets-of-jellyfish/

[18]   Fleiss A, Sarkisyan KS. A brief review of bioluminescent systems (2019). Curr Genet 2019; 65(4):

877-82.
[http://dx.doi.org/10.1007/s00294-019-00951-5] [PMID: 30850867]

[19]   Significance of bacterial bioluminescence in agriculture. Knowledge Consortium of Gujarat. Jour Sci ISSN- 2320-0006 http://kcgjournal.org/kcg/wp-content/uploads/Science/issue15/Issue15ProfKrunal& 2DrSwati&DrSanjayJha.pdf

[20]   Tampier M. Bioluminescence—the light of living things. Creation 2017; 39: 20-3.
[http://dx.doi.org/10.1007/s10661-020-08685-5]

[21]   Gouveneaux A, Flood PR, Mallefet J. Unexpected diversity of bioluminescence in planktonic worms. Luminescence 2017; 32(3): 394-400.
[http://dx.doi.org/10.1002/bio.3192] [PMID: 27545998]

[22]   Kroemer T. Colourful life under the sea from bioluminescence to ultra black.
https://www.goldbio.com/blog/post?slug=colorful-life-under-the-sea-from-bioluminescence-to-ultra-black+

[23]   Latz MI, Frank TM, Case JF. Spectral composition of bioluminescence of epipelagic organisms from the Sargasso Sea. Mar Biol 1988; 98(3): 441-6.
[http://dx.doi.org/10.1007/BF00391120]

[24]   Widder EA. Bioluminescence in the ocean: origins of biological, chemical, and ecological diversity. Science 2010; 328(5979): 704-8.
[http://dx.doi.org/10.1126/science.1174269] [PMID: 20448176]

[25]   Green AA, McElroy WD. Crystalline firefly luciferase. Biochim Biophys Acta 1956; 20(1): 170-6.
[http://dx.doi.org/10.1016/0006-3002(56)90275-X] [PMID: 13315363]

[26]   Lee J. Basic Bioluminescence Photobiology.info/LeeBasicBiolum.html

[27]   Campbell AK, Herring PJ. Imidazolopyrazine bioluminescence in copepods and other marine organisms. Mar Biol 1990; 104(2): 219-25.
[http://dx.doi.org/10.1007/BF01313261]

[28]   Luciferin Details. https://biolum.eemb.ucsb.edu/chem/detail2.html

[29]   Krill luciferin. https://pubchem.ncbi.nlm.nih.gov/compound/102018493

[30]   Vargulin. https://en.wikipedia.org/wiki/Vargulin

[31]   Gonzalez VM Jr. Synthesis, Luminescence, and applications of coelenterazine and its analogs. 2007. https://chemistry.illinois.edu/system/files/inline-files/03GonzalezFINALAbstract.pdf

[32]   Markova SV, Larionova MD, Vysotski ES. Shining Light on the Secreted Luciferases of Marine Copepods: Current Knowledge and Applications. Photochem Photobiol Sci 2019; 95(3): 705-21.
[http://dx.doi.org/10.1111/php.13077] [PMID: 30585639]

[33]   Valiadi M, Iglesias-Rodriguez D. Understanding Bioluminescence in Dinoflagellates-How Far Have We Come? Microorganisms 2013; 1(1): 3-25.
[http://dx.doi.org/10.3390/microorganisms1010003] [PMID: 27694761]

[34]   Kanie S, Komatsu M, Mitani Y. Luminescence of *Cypridina* Luciferin in the Presence of Human Plasma Alpha 1-Acid Glycoprotein. Int J Mol Sci 2020; 21(20): 7516.
[http://dx.doi.org/10.3390/ijms21207516] [PMID: 33053850]

[35]   Why do animals glow? A guide to Bioluminescence. https://oceanconservancy.org/blog/2019/08/06/animals-glow-bioluminescence/

[36]   Rodicheva EK, Vydryakova GA, Medvede SE. The IBSO catalogue of luminous bacteria cultures.
http://www.ibp.ru/collection/ibpCatalog-2017.pdf

[37]   Santhanam R, Ramanathan N, Venkataramnujam K, Jegatheesan G. Phytoplankton of the Indian Seas Daya Publishing House Delhi. 1989.

[38] Widder EA. Bioluminescence and the pelagic visual environment. Mar Freshwat Behav Physiol 2002; 35(1-2): 1-26.
[http://dx.doi.org/10.1080/10236240290025581]

[39] Biggley WH, Swift E, Buchanan RJ, Seliger HH. Stimulable and spontaneous bioluminescence in the marine dinoflagellates, *Pyrodinium bahamense, Gonyaulax polyedra*, and *Pyrocystis lunula.* J Gen Physiol 1969; 54(1): 96-122.
[http://dx.doi.org/10.1085/jgp.54.1.96] [PMID: 5792367]

[40] Esaias WE, Curl HC Jr. Effect of dinoflagellate bioluminescence on copepod ingestion rates. Limnol Oceanogr 1972; 17(6): 901-6.
[http://dx.doi.org/10.4319/lo.1972.17.6.0901]

[41] Swift E, Biggley WH, Seliger HH. Sspecies of oceanic dinoflagellates in the genera *Dissodinium* and *Pyrocystis*: interclonal and interspecific comparisons of the color and photon yield of bioluminescence. J Phycol 1973; 9(4): 420-6.

[42] Widder EA, Latz MI, Case JF. Marine bioluminescence spectra measured with an optical multichannel detection system. Biol Bull 1983; 165(3): 791-810.
[http://dx.doi.org/10.2307/1541479] [PMID: 29324013]

[43] Buskey EJ, Swift E. An encounter model to predict natural planktonic bioluminescence. Limnol Oceanogr 1990; 35(7): 1469-85.
[http://dx.doi.org/10.4319/lo.1990.35.7.1469]

[44] Drebes G. *Dissodinium pseudolunula* (Dinophyta), a parasite on copepod eggs. Br Phycol J 1978; 13(4): 319-27.
[http://dx.doi.org/10.1080/00071617800650381]

[45] *Pyrocystis fusiformis.* https://en.wikipedia.org/wiki/Pyrocystis_fusiformis

[46] Latz MI, Naueni JC, Rohr J. Bioluminescence response of four species of dinoflagellates to fully developed pipe flow. J Plankton Res 2004; 26(12): 1529-46.
[http://dx.doi.org/10.1093/plankt/fbh141]

[47] Fajardo C, Donato MD, Rodulfo H, *et al.* New Perspectives Related to the Bioluminescent System in Dinoflagellates: *Pyrocystis lunula* , a Case Study. Int J Mol Sci 21(5): 1784.2022; https://www.vims.edu/bayinfo/habs/guide/alexandrium.php

[49] Red waters in Ría de Vigo (NW Spain). Harmful Algae News 2018. www.ioc-unesco.org/hab

[50] Faust MA, Gulledge RA. Harmful Marine Dinoflagellates. http://species-identification.org/ species.php?species_group=dinoflagellates&id=60

[51] Riccardi M, Guerrin F, Fattorusso E, *et al. Gonyaulax spinifera* from the Adriatic sea: Toxin production and phylogenetic analysis. Harmful Algae 2009; 8(2): 279-90.
[http://dx.doi.org/10.1016/j.hal.2008.06.008]

[52] Lingulodinium polyedra https://en.wikipedia.org/wiki/Lingulodinium_polyedra

[53] Latz MI, Frank TM, Case J, Swift E, Bidigare R. Bioluminescence of Colonial Radiolaria in the Western Sargasso Sea. J Exp Mar Biol Ecol 1987; 109(1): 25-38.
[http://dx.doi.org/10.1016/0022-0981(87)90183-3]

[54] Herring PJ. Some features of the bioluminescence of the radiolarian *Thalassicolla sp.* Mar Biol 1979; 53(3): s213-6.
[http://dx.doi.org/10.1007/BF00952428]

[55] Latz MI, Bowlby MR, Case JF. MR, Case JF. Bioluminescence of the solitary spumellarian radiolarian, *Thalassicolla nucleata* (Huxley). J Plankton Res 1991; 13(6): 1187-202.
[http://dx.doi.org/10.1093/plankt/13.6.1187]

[56] *Tuscaridium cygneum.* https://biolum.eemb.ucsb.edu/organism/pictures/radiolarian.html

[57] Haddock SHD, Case J. Bioluminescence spectra of shallow and deep-sea gelatinous zooplankton: ctenophores, medusae and siphonophores. Mar Biol 1999; 133(3): 571-82.
[http://dx.doi.org/10.1007/s002270050497]

[58] Hunt ME, Modi CK, Aglyamova GV, Ravikant DVS, Meyer E, Matz MV. Multi-domain GFP-like proteins from two species of marine hydrozoans. Photochem Photobiol Sci 2012; 11(4): 637-44.
[http://dx.doi.org/10.1039/c1pp05238a] [PMID: 22251928]

[59] Xia NS, Luo WX, Zhang J, *et al.* Bioluminescence of *Aequorea macrodactyla*, a common jellyfish species in the East China Sea. Mar Biotechnol (NY) 2002; 4(2): 155-62.
[http://dx.doi.org/10.1007/s10126-001-0081-7] [PMID: 14961275]

[60] Widder EA, Bernstein SA, Bracher DF, *et al.* Bioluminescence in the Monterey Submarine Canyon: image analysis of video recordings from a midwater submersible. Mar Biol 1989; 100(4): 541-51.
[http://dx.doi.org/10.1007/BF00394831]

[61] Clarke GL, Conover RI, David CN, Nicol JAC. Comparative studies of luminescence in copepods and other pelagic animals. J Mar Biol Assoc U K 1962; 42(3): 541-64.
[http://dx.doi.org/10.1017/S0025315400054254]

[62] Santhanam R. Biology and Ecology of Venomous Marine Cnidarians. Springer Nature. 2020.
[http://dx.doi.org/10.1007/978-981-15-1603-0]

[63] Mizrahi GA, Shemesh E, van Ofwegen L, Tchernov D. Tchernov. D. First record of *Aequorea macrodactyla* (Cnidaria, Hydrozoa) from the Israeli coast of the eastern Mediterranean Sea, an alien species indicating invasive pathways. NeoBiota 2015; 26: 55-70.
[http://dx.doi.org/10.3897/neobiota.26.8278]

[64] Kubota S. Various distribution patterns of green fluorescence in small hydromedusae. Bull Biol Inst Kuroshio 2010; 6: 11-4.

[65] Markova SV, Burakova LP, Frank LA, Golz S, Korostileva KA, Vysotski ES. Green-fluorescent protein from the bioluminescent jellyfish *Clytia gregaria*: cDNA cloning, expression, and characterization of novel recombinant protein. Photochem Photobiol Sci 2010; 9(6): 757-65.
[http://dx.doi.org/10.1039/c0pp00023j] [PMID: 20442953]

[66] Baba K, Miyazono A, Matsuyama K, Kohno S, Kubota S. Occurrence and detrimental effects of the bivalve-inhabiting hydroid *Eutima japonica* on juveniles of the Japanese scallop *Mizuhopecten yessoensis* in Funka Bay, Japan: relationship to juvenile massive mortality in 2003. Mar Biol 2007; 151(5): 1977-87.
[http://dx.doi.org/10.1007/s00227-007-0636-x] [PMID: 30363784]

[67] Haddock SHD, Rivers TJ, Robison BH. Can coelenterates make coelenterazine? Dietary requirement for luciferin in cnidarian bioluminescence. Proc Natl Acad Sci USA 2001; 98(20): 11148-51.
[http://dx.doi.org/10.1073/pnas.201329798] [PMID. 11572972]

[68] Widder EA. Midwater bioluminescence assessment in the West Alboran Gyre l (Mediterranean Sea). Harbor Branch Oceanographic Institution RN Seward Johnson Cruise Report 1991.

[69] Mok K. Scientists Encounter Jellyfish That Looks Like "Deep Sea Fireworks". 2018.
https://www.treehugger.com/natural-sciences/halitrephes-maasi-ev-nautilus.html

[70] Cross Jellyfish. https://www.vichighmarine.ca/tag/mitrocoma-cellularia/

[71] Burakova LP, Natashin PV, Markova SV, *et al.* Mitrocomin from the jellyfish *Mitrocoma cellularia*a with deleted C-terminal tyrosine reveals a higher bioluminescence activity compared to wild type photoprotein. J Photochem Photobiol B 2016; 162: 286-97.
[http://dx.doi.org/10.1016/j.jphotobiol.2016.06.054] [PMID: 27395792]

[72] Morse VJ. The regulation and origin of bioluminescence in the hydroid Obelia. PhD dissertation Cardiff Wales: Cardiff University. 2013.

[73] Solmissus https://doris.ffessm.fr/Especes/Solmissus-albescens-Solmissus-495

[74]   Hisada M. A Study on the Photoreceptor of a Medusa, *Spirocodon saltatrix*. Jour Fac Sci Hokkaiao Univ Ser VI Zool 1956; 12: 529-33.

[75]   Calder DR, Crow GL, Ikeda S, *et al. Tima nigroannulata* (Cnidaria: Hydrozoa: Eirenidae), a New Species of Hydrozoan from Japan. Zool Sci 2021; 38(4): 370-82.
        [http://dx.doi.org/10.2108/zs210011] [PMID: 34342958]

[76]   Haddock SHD, Dunn CW, Pugh PR, Schnitzler CE. Bioluminescent and red-fluorescent lures in a deep-sea siphonophore. Science 2005; 309(5732): 263.
        [http://dx.doi.org/10.1126/science.1110441] [PMID: 16002609]

[77]   Gershwin L, Lewis M, Gowlett-Holmes K, Kloser R. The Siphonophores. Pelagic Invertebrates of South-Eastern Australia: A field reference guide. Hobart: CSIRO Marine and Atmospheric Research 2014.

[78]   Mackie GO, Pugh PR, Purcell JE. Siphonophore Biology. Adv Mar Biol 1988; 24: 97-262.
        [http://dx.doi.org/10.1016/S0065-2881(08)60074-7]

[79]   Ocean life with out sunlight. https://www.blogofazoologist.co.uk/post/ocean-life-without-sunlight

[80]   Stephanomia amphytridis https://biolum.eemb.ucsb.edu/organism/pictures/Stephanomia.html

[81]   Shimomura O, Flood PR. Luciferase of the Scyphozoan Medusa *Periphylla periphylla*. Biol Bull 1998; 194(3): 244-52.
        [http://dx.doi.org/10.2307/1543094] [PMID: 28570201]

[82]   Herring PJ, Widder EA. Bioluminescence of deep-sea coronate medusae (Cnidaria: Scyphozoa). Mar Biol 2004; 146(1): 39-51.
        [http://dx.doi.org/10.1007/s00227-004-1430-7]

[83]   Arai MA. Functional Biology of Scyphozoa. Springer Science & Business Media: Science . 2012.

[84]   Johnsen S. Bioluminescence https://oceanexplorer.noaa.gov/explorations/15biolum/background/biolum/biolum.html

[85]   Frank TM, Widder EA, Latz MI, Case JF. Dietary maintenance of bioluminescence in a deep-sea mysid. J Exp Biol 1984; 109(1): 385-9.
        [http://dx.doi.org/10.1242/jeb.109.1.385]

[86]   Woods A. Facts on the scyphozoan jellyfish *Atolla wyvillei*. https://animals.mom.com/scyphozoan-jellyfish-atolla-wyvillei-3409.html

[87]   Nikolaevich TY, Vladimirovna MO. Bioluminescence of the Black Sea Ctenophores-Aliens as an Index of their Physiological State. 2016.
        [http://dx.doi.org/10.5772/65063]

[88]   Young RE. Oceanic bioluminescence: an overview of general functions. Bull Mar Sci 1983; 33: 829-45.

[89]   Mckenzie W, Parker B. Aquatic Inverteberate Cell Culture. Scientific e-Resources 2019.

[90]   Bioluminescence in Jellyfish. https://www.reed.edu/biology/professors/srenn/pages/teaching/web_2010/mi_site/Adaptivevalue.html

[91]   Cestida. https://www.accessscience.com/content/cestida/122300

[92]   Francis WR, Shaner NC, Christianson LM, Powers ML, Haddock SHD. Occurrence of Isopenicillin-N-Synthase Homologs in Bioluminescent Ctenophores and Implications for Coelenterazine Biosynthesis. PLoS One 2015; 10(6): e0128742.
        [http://dx.doi.org/10.1371/journal.pone.0128742] [PMID: 26125183]

[93]   Haddock SHD, Moline MA, Case JF. Bioluminescence in the sea. Annu Rev Mar Sci 2010; 2(1): 443-93.
        [http://dx.doi.org/10.1146/annurev-marine-120308-081028] [PMID: 21141672]

[94]   Moving cilia create iridescence. Beroe https://asknature.org/strategy/moving-cilia-create-iridescence/

[95]   Bioluminescence. https://biolum.eemb.ucsb.edu/organism/pictures/beroe.html

[96]   Bioluminescence in jellfish https://www.reed.edu/biology/professors/srenn/pages/teaching/web_2010/mi_site/Adaptivevalue.html

[97]   Tokarev Y, Mashukova O, Sibirtsov E. Bioluminescence Characteristics Changeability of Ctenophore *Beroe ovata*. Mayer, 1912 (Beroida) in Ontogenesis. Turk J Fish Aquat Sci 2012; 12(2): 479-84.
       [http://dx.doi.org/10.4194/1303-2712-v12_2_39]

[98]   Verdes A, Gruber DF. Glowing Worms: Biological, Chemical, and Functional Diversity of Bioluminescent Annelids. Integr Comp Biol 2017; 57(1): 18-32.
       [http://dx.doi.org/10.1093/icb/icx017] [PMID: 28582579]

[99]   Dales RP. Bioluminescence in pelagic polychaetes. J Fish Res Board Can 1971; 28(10): 1487-9.
       [http://dx.doi.org/10.1139/f71-228]

[100] Osborn KJ, Haddock SHD, Pleijel F, Madin LP, Rouse GW. Deep-sea, swimming worms with luminescent "bombs". Science 2009; 325(5943): 964.
       [http://dx.doi.org/10.1126/science.1172488] [PMID: 19696343]

[101] Burnette AB, Struck TH, Halanych KM. Holopelagic *Poeobius meseres* ("Poeobiidae," Annelida) is derived from benthic flabelligerid worms. Biol Bull 2005; 208(3): 213-20.
       [http://dx.doi.org/10.2307/3593153] [PMID: 15965126]

[102] Francis WR, Powers ML, Haddock SHD. Characterization of an anthraquinone fluor from the bioluminescent, pelagic polychaete *Tomopteris*. Luminescence 2014; 29(8): 1135-40.
       [http://dx.doi.org/10.1002/bio.2671] [PMID: 24760626]

[103] Ramesh CH, Mohanraju R, Karthik P, Kutty KN. Distribution of bioluminescent polychaete larvae of *Odontosyllis* sp. in South Andaman. Indian J Geo-Mar Sci 2017; 46(4): 735-7.

[104] Gouveneaux A. Bioluminescence of Tomopteridae species (Annelida): multidisciplinary approach. PhD dissertation. Université catholique de Louvain. 2016.

[105] Brooke C. This deep sea alien worm *Tomopteris* is utterly captivating https://featuredcreature.com/this-deep-sea-alien-worm-*Tomopteris*-is-utterly-captivating/

[106] Mirza JD, Migotto AE, Yampolsky IV, de Moraes GV, Tsarkova AS, Oliveira AG. *Chaetopterus variopedatus* bioluminescence: A review of light emission within a species complex. Photochem Photobiol 2020; 96(4): 768-78.
       [http://dx.doi.org/10.1111/php.13221] [PMID: 32012290]

[107] Jyothi M, Suneetha V. Reported Bioluminescent Organisms on Land and in the Sea. Research J Pharm Tech 2017, 10(8). 1-5.
       [http://dx.doi.org/10.5958/0974-360X.2017.00640.0]

[108] Thuesen EV, Goetz FE, Haddock SHD. Bioluminescent organs of two deep-sea arrow worms, *Eukrohnia fowleri* and *Caecosagitta macrocephala*, with further observations on Bioluminescence in chaetognaths. Biol Bull 2010; 219(2): 100-11.
       [http://dx.doi.org/10.1086/BBLv219n2p100] [PMID: 20972255]

[109] Raymond JA, DeVries AL. Bioluminescence in McMurdo Sound, Antarctica. Limnol Oceanogr 1976; 21(4): 599-602.
       [http://dx.doi.org/10.4319/lo.1976.21.4.0599]

[110] Morin JG. Luminaries of the reef: The history of luminescent ostracods and their courtship displays in the Caribbean. J Crustac Biol 2019; 39(3): 227-43.
       [http://dx.doi.org/10.1093/jcbiol/ruz009]

[111] Mitani Y, Oshima Y, Mitsuda N, *et al.* Efficient production of glycosylated Cypridina luciferase using plant cells. Protein Expr Purif 2017; 133: 102-9.

[http://dx.doi.org/10.1016/j.pep.2017.03.008] [PMID: 28288897]

[112]   Tsuji FI, Lynch RV III, Haneda Y. Studies on the bioluminescence of the marine ostracod crustacean *Cypridina serrata*. Biol Bull 1970; 139(2): 386-401.
[http://dx.doi.org/10.2307/1540092] [PMID: 29332462]

[113]   Ogoh K, Ohmiya Y. Biogeography of luminous marine ostracod driven irreversibly by the Japan current. Mol Biol Evol 2005; 22(7): 1543-5.
[http://dx.doi.org/10.1093/molbev/msi155] [PMID: 15858206]

[114]   Torres E, Cohen AC. Vargula morini, a New Species of Bioluminescent Ostracode (Myodocopida: Cypridinidae) from Belize and an Associated Copepod (Copepoda: Siphonostomatoida: Nicothoidae). J Crustac Biol 2005; 25(1): 11-24.
[http://dx.doi.org/10.1651/C-2455]

[115]   Heger A, King NJ, Wigham BD, *et al*. Benthic bioluminescence in the bathyal North East Atlantic: luminescent responses of *Vargula norvegica* (Ostracoda: Myodocopida) to predation by the deep-water eel (Synaphobranchus kaupii). Mar Biol 2007; 151(4): 1471-8.
[http://dx.doi.org/10.1007/s00227-006-0587-7]

[116]   Kornicker LS, King CE. A New Species of Luminescent Ostracoda from Jamaica, West Indies. Micropaleontology 1965; 11(1): 105-10.
[http://dx.doi.org/10.2307/1484821]

[117]   Marine Planktonic Ostracods. *Conchoecia spinifera*. http://species-identification.org/species.php? species_group=ostracods&id=98

[118]   Zooplankton Guide *Conchoecia magna*. https://scripps.ucsd.edu/zooplanktonguide/species/conchoecia-magna

[119]   Lapota D, Geiger ML, Stiffey AV, Rosenberger DE, Young DK. Correlations of planktonic bioluminescence with other oceanographic parameters from a Norwegian fjord. Mar Ecol Prog Ser 1989; 55: 217-27.
[http://dx.doi.org/10.3354/meps055217]

[120]   Oba Y, Tsuduki H, Kato S, Ojika M, Inouye S. Identification of the luciferin-luciferase system and quantification of coelenterazine by mass spectrometry in the deep-sea luminous ostracod *Conchoecia pseudodiscophora*. ChemBioChem 2004; 5(11): 1495-9.
[http://dx.doi.org/10.1002/cbic.200400102] [PMID: 15515099]

[121]   Batchelder HP, Swift E. Bioluminescent Potential and Variability in Some Sargasso Sea Planktonic Halocyprid Ostracods. J Crustac Biol 1988; 8(4): 520-3.
[http://dx.doi.org/10.2307/1548687]

[122]   Santhanam R, Srinivasan A. A Manual of Marine Zooplankton. Oxford & IBH Publishing. New Dehli. 1994.

[123]   Takenaka Y, Yamaguchi A, Tsuruoka N, Torimura M, Gojobori T, Shigeri Y. Evolution of bioluminescence in marine planktonic copepods. Mol Biol Evol 2012; 29(6): 1669-81.
[http://dx.doi.org/10.1093/molbev/mss009] [PMID: 22319154]

[124]   Lapota D, Losee JR. Observations of bioluminescence in marine plankton from the Sea of Cortez. Ecology 1984; 77(3): 209-39.

[125]   Masuda H, Takenaka Y, Yamaguchi A, Nishikawa S, Mizuno H. A novel yellowish-green fluorescent protein from the marine copepod, *Chiridius poppei*, and its use as a reporter protein in HeLa cells. Gene 2006; 372(1): 18-25.
[http://dx.doi.org/10.1016/j.gene.2005.11.031] [PMID: 16481130]

[126]   Lapota D, Losee JR, Geiger ML. Bioluminescence displays induced by pulsed light. Limnol Oceanogr 1986; 31(4): 887-9.
[http://dx.doi.org/10.4319/lo.1986.31.4.0887]

[127]   Bannister NJ, Herrring PJ. Distribution and Structure of Luminous Cells In four Marine Copepods. J Mar Biol Assoc U K 1989; 69(3): 523-33.
[http://dx.doi.org/10.1017/S0025315400030939]

[128]   Larionova MD, Markova SV, Tikunova NV, Vysotski ES. The Smallest Isoform of *Metridia longa* Luciferase as a Fusion Partner for Hybrid Proteins. Int J Mol Sci 2020; 21: 497.
[http://dx.doi.org/10.3354/meps094297]

[129]   Bowlby MR, Widder EA, Case JF. Disparate forms of bioluminescence from the amphipods *Cyphocaris faurei, Scina crassicornis* and *S. borealis*. Mar Biol 1991; 108(2): 247-53.
[http://dx.doi.org/10.1007/BF01344339]

[130]   Johnsen S. The red and the black: bioluminescence and the color of animals in the deep sea. Integr Comp Biol 2005; 45(2): 234-46.
[http://dx.doi.org/10.1093/icb/45.2.234] [PMID: 21676767]

[131]   Neognathophausia   gigas
https://inverts.wallawalla.edu/Arthropoda/Crustacea/Malacostraca/Eumalacostraca/Peracarida/Lophog astrida/Neognathophausia_gigas.html

[132]   Shimomura O, Yampolsk IV. Bioluminescence: Chemical Principles And Methods. 3rd ed., World Scientific 2019.
[http://dx.doi.org/10.1142/11180]

[133]   Nakamura H, Musicki B, Kishi Y, Shimomura O. Structure of the light emitter in krill (*Euphausia pacifica*) bioluminescence. J Am Chem Soc 1988; 110(8): 2683-5.
[http://dx.doi.org/10.1021/ja00216a070]

[134]   Antarctic krill.. https://en.wikipedia.org/wiki/Antarctic_krill

[135]   Kay RH. Light-stimulated and light-inhibited bioluminescence of the euphausiid Meganyctiphanes norvegica (G.O. Sars). https://royalsocietypublishing.org/doi/pdf/10.1098/rspb.1965.0044

[136]   Doyle JD, Kay RH. Some Studies on the Bioluminescence of the Euphausiids, *Meganyctiphanes norvegica* and *Thysanoessa raschii*. J Mar Biol Assoc U K 1967; 47(3): 555-63.
[http://dx.doi.org/10.1017/S0025315400035189]

[137]   Herring PJ. The biology of the deep ocean. Peter herring. Oxford University Press. 2002.

[138]   *Nyctiphanes-australis*. https://www.imas.utas.edu.au/zooplankton/image-key/malacostraca/euphausia cea/nyctiphanes-australis

[139]   van Couwelaar M. Zooplankton and Micronekton of the North Sea *Stylocheiron abbreviatum* http://species-identification.org/species.php?species_group=zmns&id=458

[140]   Lucifer (prawn) https://en.wikipedia.org/wiki/Lucifer_(prawn)

[141]   Galt CP, Sykes PF. Sites of bioluminescence in the appendicularians *Oikopleura dioica* and *O. labradoriensis* (Urochordata: Larvacea). Mar Biol 1983; 77(2): 155-9.
[http://dx.doi.org/10.1007/BF00396313]

[142]   Galt CP, Grober MS, Sykes PF. Taxonomic correlates of bioluminescence among appendicularians (Urochordata: Larvacea). Biol Bull 1985; 168(1): 125-34.
[http://dx.doi.org/10.2307/1541178]

[143]   Robison BH, Raskoff KA, Sherlock RE. Ecological substrate in midwater: *Doliolula equus*, a new mesopelagic tunicate. J Mar Biol Assoc UK 2005; 85(3): 655-63.
[http://dx.doi.org/10.1017/S0025315405011586]

[144]   Bowlby MR, Widder EA, Case JF. Patterns of Stimulated Bioluminescence in Two Pyrosomes (Tunicata: Pyrosomatidae). Biol Bull 1990; 179(3): 340-50.
[http://dx.doi.org/10.2307/1542326] [PMID: 29314963]

[145]   Tessler M, Gaffney JP, Oliveira AG, *et al.* A putative chordate luciferase from a cosmopolitan tunicate

indicates convergent bioluminescence evolution across phyla. Sci Rep 2020; 10(1): 17724.
[http://dx.doi.org/10.1038/s41598-020-73446-w] [PMID: 33082360]

[146] Unicorn of the Sea- Incredible Pyrosome deep-sea giant worm https://wordlesstech.com/unicorn-o-
-the-sea-incredible-pyrosome-deep-sea-giant-worm/

[147] Melina R. Gallery: Eye-Catching Bioluminescent Wonders. https://www.livescience.com/14865-
bioluminescent-creatures-gallery/2.html

[148] Soest RWMV. Taxonomy of the subfamily Cyclosalpinae Yount, 1954 (Tunicata, Thaliacea), with
descriptions of two new species. Beaufortia 1974; 22(288): 17-55.

[149] Dybas CL. Illuminating New Biomedical Discoveries: Bioluminescent, biofluorescent species glow
with promise. Bioscience 2019; 69(7): 487-95.
[http://dx.doi.org/10.1093/biosci/biz052]

[150] https://www.cancer.gov/publications/dictionaries/cancer-terms/def/car-t-cell-therapy

[151] Ramesh Ch, Mohanraju R. A Review on Bioluminescence and its Applications Int J Lum App. 2015;
5(1): 45-6.

[152] Syed AJ, Anderson JC. Applications of bioluminescence in biotechnology and beyond. Chem Soc Rev
2021; 50(9): 5668-705.
[http://dx.doi.org/10.1039/D0CS01492C] [PMID: 33735357]

[153] https://slidetodoc.com/marine-biotechnology-lab-bioluminescence-bioluminescence-is-the-ability/

SUBJECT INDEX

A

Abylopsis 50, 51, 84, 85
   eschscholtzii 51, 84
   tetragona 50, 51, 85
Acartia claui 168
Acrocalanus longicornis 168
Acrosphaera murrayana 44, 45
Acute 201, 228
   myelogenous leukemia 228
   rostrum 201
Aegina citrea 50, 51, 53, 54
Aeginidae aegina 51
Aeginura grimaldii 51, 54
Alciopids 145
Alciopina 142
Aldehydes, long-chain 12, 17
Alexandrium 20, 32
   acatenella 20
   affine 20
   catenella 20
   fraterculus 20
   fundyense 20
   minutum 32
   ostenfeldii 20
   tamarense 20
Algae 19, 25, 48, 49
   photosynthetic unicellular 48
Allergic conditions 172
Alzheimer's disease 7
Amino acids 14, 16, 123
   tryptophan 14, 16
Amphicaryon acaule 51
Amphipods 5, 158, 181, 183
   bioluminescent marine 181
   luminous marine 183
Annelids, luminescent planktonic marine 144
Antapical 31, 38, 39
   horns 31, 38, 39
   margin 39
Antarctica 102
Anterior cirri 148, 151
Apical 24, 80, 89

diverticulum 89
groove 24
mesoglea 80
Apical horn 31, 33, 38, 39
   robust 39
Arctic 176
   ocean diversity 176
Assay 228, 229
   bioluminescence ATP 229
Atlantic 58, 60, 85, 89, 90, 92, 94, 95, 98, 125,
   131, 133, 139, 215, 216, 217
   coast 98
   ocean 58, 60, 85, 89, 90, 92, 94, 95, 125,
   131, 133, 139, 215, 216, 217
Atolla vanhoeffeni 104, 105, 106
ATP 15, 227, 228
   bioluminescence assay 228
Aulacoctena acuminata 118, 119

B

Bacteria 1, 3, 4, 7, 12, 15, 17, 25, 213, 228
   halophilous 17
   infectious 228
   luminous 17, 213
Bathocyroe fosteri 117, 118, 119, 120
Bathyctena chuni 117, 118, 120, 121
Bentheuphausia ambylops 185
Benthic 4, 158
   crabs 158
   environment 4
Beroe 118, 137, 138, 139
   cucumis 118, 137
   forskalii 118, 138, 139
   gracilis 118, 138, 139
Bioluminescence 1, 9, 15, 16, 18, 36, 42, 157,
   134, 220, 228, 229
   chemistryis 9
   coelenterazine-based 157
   detecting oceanographic-scale 229
   dinoflagellate 42
   intensity 134
   luciferin-based 18

V

Velamen parallelum 118, 119, 136
Vibrio 17
  fischeri 17
  harveyi 17
Voluminous stomodaeum 130

W

Warts 63, 112
  marginal 63
Waters 2, 3, 5, 8, 29, 34, 57, 60, 64, 106, 112,
      114, 116, 122, 133, 152, 155, 158, 162
  brackish 122, 133
  estuarine 23
  mesopelagic 152
  nearshore 34, 57
  offshore 60
  tropical ocean 29
  tropical surface 155
Wavelengths 49, 72, 184
  blue-green 49, 184
  emission 72
Waves 19, 35, 60, 95, 102, 103, 111
  splashing 19
Wikimedia 54, 55, 65, 66, 94, 95, 96, 98, 99,
      106, 107, 109, 110, 111, 112
Wikipedia 26, 27, 30, 33, 35, 45, 58, 134, 191

www.ingramcontent.com/pod-product-compliance
Lightning Source LLC
Chambersburg PA
CBHW050821220326
41598CB00006B/283